Praise for *Eaarth*

"'Eaarth' is the name McKibben has decided to assign both to his new book and to the planet formerly known as Earth. His point is a fresh one that brings the reader uncomfortably close to climate change. Earth with one 'a,' according to McKibben, no longer exists. We have carbonized it out of existence. Two-a Eaarth is now our home." —*The New York Times Book Review*

"A passionate appeal . . . McKibben's engaging and persuasive book will add greatly to the sense of urgency. It will add realism to the case for strong adaptation to the changes that our past and current actions are bringing to our natural world."

—*The New York Review of Books*

"A valuable slice of acid-tongued reality."

—*San Francisco Chronicle*

"If one book can help, this is it." —*Winnipeg Free Press*

"This book must be read and his message must be understood clearly in Congress and in the streets. Indeed, throughout the world." —*The Capitol Times* (Madison, Wisconsin)

"Sounds a clarion at a time when the findings of climate scientists have been all but drowned out by skeptics and right-wing bombast. McKibben, however, does not doubt that facts will trump ideology. . . . McKibben is an eloquent advocate."

—*The Oregonian*

"With clarity, eloquence, deep knowledge, and even deeper compassion for both planet and people, Bill McKibben guides us to the brink of a new, uncharted era. This monumental book, probably his greatest, may restore your faith in the future, with us in it."

—Alan Weisman, author of *The World Without Us*

"The terrifying premise with which this book begins is that we have, as in the old science fiction films and tales of half a century ago, landed on a harsh and unpredictable planet, all six billion of us. Climate change is already here, but Bill McKibben doesn't stop with the bad news. He tours the best responses that are also already here, and these visions of a practical scientific solution are also sketches of a better, richer, more democratic civil society and everyday life. *Eaarth* is an astonishingly important book that will knock you down and pick you up."

—Rebecca Solnit, author of *A Paradise Built in Hell* and *Hope in the Dark*

"Bill McKibben foresaw 'the end of nature' very early on, and in this new book he blazes a path to help preserve nature's greatest treasures."

—James E. Hansen, director of NASA Goddard Institute for Space Studies

"Bill McKibben is the most effective environmental activist of our age. Anyone interested in making a difference to our world can learn from him."

—Tim Flannery, author of *The Weather Makers* and *The Eternal Frontier*

ALSO BY BILL MCKIBBEN

The Bill McKibben Reader

Fight Global Warming Now

Deep Economy

Wandering Home

Enough

Long Distance

Hundred Dollar Holiday

Maybe One

The Comforting Whirlwind

Hope, Human and Wild

The Age of Missing Information

The End of Nature

Eaarth

MAKING A LIFE ON A TOUGH NEW PLANET

BILL McKIBBEN

St. Martin's Griffin
New York

Text stock contains 20% post-consumer waste
recycled fiber

For information, address St. Martin's Press,
175 Fifth Avenue, New York, N.Y. 10010.

www.stmartins.com

Designed by Kelly S. Too

The Library of Congress has cataloged the Henry Holt edition as follows:

McKibben, Bill.
Eaarth: making a life on a tough new planet / Bill McKibben.—1st ed.
p. cm.
Includes bibliographical references and index.
ISBN 978-0-8050-9056-7
1. Climatic changes. 2. Global warming. 3. Greenhouse effect, Atmospheric.
4. Environmental degradation. 5. Nature—Effect of human beings on. I. Title.
QC981.8.C5M3895 2010
304.2—dc22 2009030040

ISBN 978-0-312-54119-4 (trade paperback)

Originally published in hardcover in 2010 by Times Books,
an imprint of Henry Holt and Company

First St. Martin's Griffin Edition: March 2011

10 9 8 7 6 5 4 3 2 1

For Phil Aroneanu, Will Bates, Kelly Blynn, May Boeve, Jamie Henn, Jeremy Osborn, Jon Warnow, and the thousands and thousands of people who work with us at 350.org

• CONTENTS •

PREFACE

I'm writing these words on a gorgeous spring afternoon, perched on the bank of a brook high along the spine of the Green Mountains, a mile or so from my home in the Vermont mountain town of Ripton. The creek burbles along, the picture of a placid mountain stream, but a few feet away there's a scene of real violence—a deep gash through the woods where a flood last summer ripped away many cubic feet of tree and rock and soil and drove it downstream through the center of the village. Before the afternoon was out, the only paved road into town had been demolished by the rushing water, a string of bridges lay in ruins, and the governor was trying to reach the area by helicopter.

Twenty years ago, in 1989, I wrote the first book for a general audience about global warming, which in those days we called the "greenhouse effect." That book, *The End of Nature,* was mainly a philosophical argument. It was too early to see the practical effects of climate change but not too early to *feel* them; in the most widely excerpted passage of the book, I described walking down a different river, near my then-home sixty miles away, in New York's Adirondack Mountains. Merely knowing that we'd begun to alter the climate meant that the water flowing in that creek had a different, lesser meaning. "Instead of a world where rain had an independent and mysterious existence, the rain had become a subset of human activity," I wrote. "The rain bore a brand; it was a steer, not a deer."

Now, that sadness has turned into a sharper-edged fear. Walking along this river today, you don't need to imagine a damned thing—the evidence of destruction is all too obvious. Much more quickly than we would have guessed in the late 1980s, global warming has dramatically altered, among many other things, hydrological cycles. One of the key facts of the twenty-first century turns out to be that warm air holds more water vapor than cold: in arid areas this means increased evaporation and hence drought. And once that water is in the atmosphere, it will come down, which in moist areas like Vermont means increased deluge and flood. Total rainfall across our continent is up 7 percent,[1] and that huge change is accelerating. Worse, more and more of it comes in downpours.[2] Not gentle rain but damaging gully washers: across the planet, flood damage is increasing by 5 percent a year.[3] Data show dramatic increases—20 percent or more—in the most extreme weather events across the eastern United States, the kind of storms that drop many inches of rain in a single day.[4] Vermont saw three flood emergencies in the 1960s, two in the 1970s, three in the 1980s—and ten in the 1990s and ten so far in the first decade of the new century.

In our Vermont town, in the summer of 2008, we had what may have been the two largest rainstorms in our history about six weeks apart. The second and worse storm, on the morning of August 6, dropped at least six inches of rain in three hours up on the steep slopes of the mountains. Those forests are mostly intact, with only light logging to disturb them—but that was far too much water for the woods to absorb. One of my neighbors,

Amy Sheldon, is a river researcher, and she was walking through the mountains with me one recent day, imagining the floods on that August morning. "You would have seen streams changing violently like that," she said, snapping her fingers. "A matter of minutes." A year later the signs persisted: streambeds gouged down to bedrock, culverts obliterated, groves of trees laid to jackstraws.

Our town of barely more than five hundred people has been coping with the damage ever since. We passed a $400,000 bond to pay for our share of the damage to town roads and culverts. (The total cost was in the millions, most of it paid by the state and federal governments.) Now we're paying more to line the creek with a seven-hundred-foot-long wall of huge boulders—riprap, it's called—where it passes through the center of town, a scheme that may save a few houses for a few years, but which will speed up the water and cause even more erosion downstream. There's a complicated equation for how wide a stream will be, given its grade and geology; Sheldon showed it to me as we reclined on rocks by the riverbank. It mathematically defines streams as we have known them, sets an upper limit to their size. You could use it to plan for the future, so you could know where to build and where to let well enough alone. But none of that planning works if it suddenly rains harder and faster than it has ever rained before, and that's exactly what's now happening. It's raining harder and evaporating faster; seas are rising and ice is melting, melting far more quickly than we once expected. The first point of this book is simple: global warming is no longer a philosophical threat, no longer a future threat, *no longer a threat at all*. It's our reality. We've changed the planet, changed it in large and fundamental ways. And these changes are far, far more evident in the toughest parts of the globe, where climate change is already wrecking thousands of lives daily. In

July 2009, Oxfam released an epic report, "Suffering the Science," which concluded that even if we now adapted "the smartest possible curbs" on carbon emissions, "the prospects are very bleak for hundreds of millions of people, most of them among the world's poorest."[5]

And so this book will be, by necessity, less philosophical than its predecessor. We need now to understand the world we've created, and consider—urgently—how to live in it. We can't simply keep stacking boulders against the change that's coming on every front; we'll need to figure out what parts of our lives and our ideologies we must abandon so that we can protect the core of our societies and civilizations. There's nothing airy or speculative about this conversation; it's got to be uncomfortable, staccato, direct.

Which doesn't mean that the change we must make—or the world on the other side—will be without its comforts or beauties. Reality always comes with beauty, sometimes more than fantasy, and the end of this book will suggest where those beauties lie. But hope has to be real. It can't be a hope that the scientists will turn out to be wrong, or that President Barack Obama can somehow fix everything. Obama can help—but precisely to the degree he's willing to embrace reality, to understand that we live on the world we live on, not the one we might wish for. Maturity is not the opposite of hope; it's what makes hope possible.

The need for that kind of maturity became painfully clear in the last days of 2009, as I was doing the final revisions for this book. Many people had invested great hope that the Copenhagen conference would mark a turning point in the climate change debate. If it did, it was a turning point for the worse, with the richest and most powerful countries making it abundantly clear that they weren't going to take strong steps to address the crisis before us. They looked the poorest and most vulnerable nations straight in the eye, and then they looked away and concluded a face-saving accord with no targets or timetables. To see hope dashed is never pleasant. In the early morning hours after President Obama jetted back to Washington, a group of young protesters gathered at the metro station outside the conference hall in Copenhagen. *It's our future you decide*, they chanted.

My only real fear is that the reality described in this book, and increasingly evident in the world around us, will be for some an excuse to give up. We need just the opposite—increased engagement. Some of that engagement will be local: building the kind of communities and economies that can withstand what's coming. And some of it must be global: we must step up the fight to keep climate change from getting even more powerfully out of control, and to try to protect those people most at risk, who are almost always those who have done the least to cause the problem. I've spent much of the last two decades in that fight, most recently helping lead 350.org, a huge grassroots global effort to force dramatic action. It's true that we've lost that fight, insofar as our goal was to preserve the world we were born into. That's not the world we live on any longer, and there's no use pretending otherwise.

But damage is always relative. So far we've increased global temperatures about a degree, and it's caused the massive change chronicled in chapter 1. That's not going to go away. But if we don't stop pouring more carbon into the atmosphere, the temperature will simply keep rising, right past the point where *any* kind of adaptation will prove impossible. I have dedicated this book to my closest colleagues in this battle, my crew at 350.org, with the pledge that we'll keep battling. We have no other choice.

Eaarth

·1·

A NEW WORLD

Imagine we live on a planet. Not our cozy, taken-for-granted earth, but a planet, a real one, with melting poles and dying forests and a heaving, corrosive sea, raked by winds, strafed by storms, scorched by heat. An inhospitable place.

It's hard. For the ten thousand years that constitute human civilization, we've existed in the sweetest of sweet spots. The temperature has barely budged; globally averaged, it's swung in the narrowest of ranges, between fifty-eight and sixty degrees Fahrenheit. That's warm enough that the ice sheets retreated from the centers of our continents so we could grow grain, but cold enough that mountain glaciers provided drinking and irrigation water to those plains and valleys year-round; it was the "correct" temperature for the marvelously diverse planet that seems right to us. And every aspect of our civilization reflects that particular world. We built our great cities next to seas that have remained tame and level, or at altitudes high enough that disease-bearing mosquitoes could not overwinter. We refined the farming that has swelled our numbers to take full advantage of that predictable heat and rainfall; our rice and corn and wheat can't imagine another earth either. Occasionally, in one place or another, there's an abrupt departure from the norm—a hurricane, a drought, a freeze. But our very language reflects their rarity: freak storms, disturbances.

In December 1968 we got the first real view of that stable, secure place. *Apollo 8* was orbiting the moon, the astronauts busy photographing possible landing zones for the missions that would follow. On the fourth orbit, Commander Frank Borman decided to roll the craft away from the moon and tilt its windows toward the horizon—he needed a navigational fix. What he got, instead, was a sudden view of the earth, rising. "Oh my God," he said. "Here's the earth coming up." Crew member Bill Anders grabbed a camera and took the photograph that became the iconic image perhaps of all time. "Earthrise," as it was eventually known, that picture of a blue-and-white marble floating amid the vast backdrop of space, set against the barren edge of the lifeless moon.[1] Borman said later that it was "the most beautiful, heart-catching sight of my life, one that sent a torrent of nostalgia, of sheer homesickness, surging through me. It was the only thing in space that had any color to it. Everything else was simply black or white. But not the earth."[2] The third member of the crew, Jim Lovell, put it more simply: the earth, he said, suddenly appeared as "a grand oasis."

But we no longer live on that planet. In the four decades since, that earth has changed in profound ways, ways that have already taken us out of the sweet spot where humans so long thrived. We're every day less the oasis and more the desert. The world hasn't ended, but the world as we know it has—even if we don't quite know it yet. We imagine we still live back on that old planet, that the disturbances we see around us are the old random and freakish kind. But they're not. It's a different place. A different planet. It needs a new name. Eaarth. Or Monnde, or

Tierrre, Errde, оккучивать. It still looks familiar enough—we're still the third rock out from the sun, still three-quarters water. Gravity still pertains; we're still earth*like*. But it's odd enough to constantly remind us how profoundly we've altered the only place we've ever known. I am aware, of course, that the earth changes constantly, and that occasionally it changes wildly, as when an asteroid strikes or an ice age relaxes its grip. This is one of those rare moments, the start of a change far larger and more thoroughgoing than anything we can read in the records of man, on a par with the biggest dangers we can read in the records of rock and ice.

Consider the veins of cloud that streak and mottle the earth in that glorious snapshot from space. So far humans, by burning fossil fuel, have raised the temperature of the planet nearly a degree Celsius (more than a degree and a half Fahrenheit). A NASA study in December 2008 found that warming on that scale was enough to trigger a 45 percent increase in thunderheads above the ocean, breeding the spectacular anvil-headed clouds that can rise five miles above the sea, generating "supercells" with torrents of rain and hail.[3] In fact, total global rainfall is now increasing 1.5 percent a decade.[4] Larger storms over land now create more lightning; every degree Celsius brings about 6 percent more lightning, according to the climate scientist Amanda Staudt. In just one day in June 2008, lightning sparked 1,700 different fires across California, burning a million acres and setting a new state record. These blazes burned on the new earth, not the old one. "We are in the mega-fire era," said Ken Frederick, a spokesman for the federal government.[5] And that smoke and flame, of course, were visible from space—indeed anyone with an Internet connection could watch the video feed from the space shuttle *Endeavour* as it circled above the towering plumes in the Santa Barbara hills.

Or consider the white and frozen top of the planet. Arctic ice has been melting slowly for two decades as temperatures have climbed, but in the summer of 2007 that gradual thaw suddenly accelerated. By the time the long Arctic night finally descended in October, there was 22 percent less sea ice than had ever been observed before, and more than 40 percent less than the year that the Apollo capsule took its picture. The Arctic ice cap was 1.1 million square miles smaller than ever in recorded history, reduced by an area twelve times the size of Great Britain.[6] The summers of 2008 and 2009 saw a virtual repeat of the epic melt; in 2008 both the Northwest and Northeast passages opened for the first time in human history. The first commercial ship to make the voyage through the newly opened straits, the MV *Camilla Desgagnes*, had an icebreaker on standby in case it ran into trouble, but the captain reported, "I didn't see one cube of ice."[7]

This is not some mere passing change; this is the earth shifting. In December 2008, scientists from the National Sea Ice Data Center said the increased melting of Arctic ice was accumulating heat in the oceans, and that this so-called Arctic amplification now penetrated 1,500 kilometers inland. In August 2009, scientists reported that lightning strikes in the Arctic had increased twentyfold, igniting some of the first tundra fires ever observed.[8] According to the center's Mark Serreze, the new data are "reinforcing the notion that the Arctic ice is in its death spiral."[9] That is, within a decade or two, a summertime spacecraft pointing its camera at the North Pole would see nothing but open ocean. There'd be ice left on Greenland—but much less ice. Between 2003 and 2008, more than a trillion tons of the island's ice melted, an area ten times the size of Manhattan. "We now know that the climate doesn't have to warm any more for Greenland to continue losing ice," explained Jason Box, a geography professor at Ohio State University. "It has probably passed the point where it could maintain the mass of ice that we remember."[10] And if the spacecraft pointed its camera at the South Pole? On the last day of 2008, the *Economist* reported that temperatures on the Antarctic Peninsula were rising faster than anywhere else on earth, and that the West Antarctic was losing ice 75 percent faster than just a decade before.[11]

Don't let your eyes glaze over at this parade of statistics (and so many more to follow). These should come as body blows, as mortar barrages, as sickening thuds. The Holocene is staggered, the only world that humans have known is suddenly reeling. I am not describing what will happen if we don't take action, or warning of some future threat. This is the *current* inventory: more thunder, more lightning, less ice. Name a major feature of the earth's surface and you'll find massive change.

For instance: a U.S. government team studying the tropics recently concluded that by the standard meteorological definition, they have expanded more than two degrees of latitude north and south since 1980—"a further 8.5 million square miles of the Earth are now experiencing a tropical climate." As the tropics expand, they push the dry subtropics ahead of them, north and south, with "grave implications for many millions of people" in these newly arid regions. In Australia, for instance, "westerly winds bringing much needed rain" are "likely to be pushed further south, dumping their water over open ocean rather than on land."[12] Indeed, by early 2008 half of Australia was in drought, and forecasters were calling it the new normal. "The inflows of the past will never return," the executive director of the Water Services Association of Australia told reporters. "We are trying to avoid the term 'drought' and saying this is the new reality."[13] They are trying to avoid the term *drought* because it implies the condition may someday *end*. The government warned in 2007 that "exceptionally hot years," which used to happen once a quarter century, would now "occur every one or two years."[14] The brushfires ignited by drought on this scale claimed hundreds of Australian lives in early 2009; four-story-high walls of flame "raced across the land like speeding trains," according to news reports. The country's prime minister visited the scene of the worst blazes. "Hell and its fury have visited the good people of Victoria," he said.[15]

And such hell is not confined to the antipodes. By the end of 2008 hydrologists in the United States were predicting that drought across the American Southwest had become a "permanent condition."[16] There was a 50 percent chance that Lake Mead, which backs up on the Colorado River behind Hoover Dam, could run dry by 2021.[17] (When that happens, as the head of the Southern Nevada Water Authority put it, "you cut off supply to the fifth largest economy in the world," spread across the American West.)[18] But the damage is already happening: researchers calculate that the new aridity and heat have led to reductions in wheat, corn, and barley yields of about 40 million tons a year.[19] The dryness keeps spreading. In early 2009 drought wracked northern China, the country's main wheat belt. Rain didn't fall for more than a hundred days, a modern record.[20] The news was much the same in India, in southern Brazil, and in Argentina, where wheat production in 2009 was the lowest in twenty years.[21] Across the planet, rivers are drying up. A massive 2009 study looked at streamflows on 925 of the world's largest rivers from 1948 to 2004 and found that twice as many were falling as rising. "During the life span of the study, fresh water discharge into the Pacific Ocean fell by about six percent—or roughly the annual volume of the Mississippi," it reported.[22]

From the flatlands to the highest peaks. The great glaciologist Lonnie Thompson, drilling cores on a huge Tibetan glacier in

2008, found something odd. Or rather, didn't find: one of the usual marker layers in any ice core, the radioactive particles that fell out from the atomic tests of the 1960s, were missing. The glacier had melted back through that history, wiped it away. A new Nepalese study found temperatures rising a tenth of a degree Fahrenheit annually in the Himalayas.[23] That would be a degree every *decade* in a world where the mercury barely budged for ten millennia. A long-standing claim that Himalayan glaciers might disappear by 2035 has been discredited, but across the region the great ice sheets are already shrinking fast: photos from the base of Mount Everest show that three hundred vertical feet of ice—a mass as tall as the Statue of Liberty—have melted since the Mallory expedition took the first photographs of the region in 1921.[24] But already, while there's still some glacier left, the new heat is flustering people. The rhododendrons that dominate Himalayan hillsides are in some places blooming forty-five days ahead of schedule, wrecking the annual spring flower festival and "creating confusion among folk artists."[25] The same kind of confusion is gripping mountaineers; one experienced high-altitude guide recently reported abandoning some mountains he'd climbed for years because "of the melting of the ice that acts as a glue, literally holding the mountains together."[26]

It's not just the Himalayas. In the spring of 2009, researchers arriving in Bolivia found that the eighteen-thousand-year-old Chacaltaya Glacier is "gone, completely melted away as of some sad, undetermined moment early this year." Once the highest ski run in the world, it now is nothing but rocks and mud.[27] But it's not the loss of a ski run that really matters. These glaciers are the reservoirs for entire continents, watering the billions of people who have settled downstream precisely because they guaranteed a steady supply. "When the glaciers are gone, they are gone. What does a place like Lima do?" asked Tim Barnett, a climate scientist at Scripps Oceanographic Institute. "In northwest China there are 300 million people relying on snowmelt for water supply. There's no way to replace it until the next ice age."[28]

When I read these accounts, I flash back to a tiny village, remote even by Tibetan standards, where I visited a few years ago. A gangly young man guided me a mile up a riverbank for a view of the enormous glacier whose snout towered over the valley. A black rock the size of an apartment tower stuck out from the middle of the wall of ice. My guide said it had appeared only the year before and now grew larger daily as its dark surface absorbed the sun's heat. We were a hundred miles from a school, far from TV; no one in the village was literate. So out of curiosity I asked the young man: "Why is it melting?" I don't know what I expected—some story about angry gods? He looked at me as if I was visiting from the planet Moron.

"Global warming," he said. "Too many factories." No confusion there. We hiked back to his hut and shook hands. I climbed into the Land Cruiser, which took me to the airplane. And so forth.

Or consider the ocean, that three-fourths of the planet that we usually don't consider. Different? One hundred eleven hurricanes formed in the tropical Atlantic between 1995 and 2008, a rise of 75 percent over the previous thirteen years. They're stronger, stranger. "Storms are not just making landfall and going away like they did in the past," said a researcher at the National Center for Atmospheric Research. "Somehow these storms are able to live longer today." In the summer of 2008, he added, "meteorologists watched in amazement as Tropical Storm Fay crisscrossed Florida a record-breaking four times" before it finally broke up; Hurricane Gustav carried its hurricane force winds all the way to Baton Rouge, a hundred miles inland, surprising the evacuees who had fled there from the coast.[29] In the last half decade we've seen the earliest-forming Category 5 hurricane ever recorded (Emily, 2005) and the first January tropical cyclone (Zeta, 2006), the first known tropical cyclone in the South Atlantic (Catarina, 2004), and the first known tropical storm ever to strike Spain (Vince, 2005). The hurricane season of 2008 was the only one on record in the Atlantic that featured major hurricanes in five separate months, from Bertha (July) to Paloma (November). And elsewhere? "The increase in ocean temperatures," according to one study, "has led Bangladesh to encounter more than twelve storm warnings per year when the previous average was three." A succession of typhoons hit the country in 2006, inundating two-thirds of the nation; a year later Cyclone Sidr killed three thousand.[30] In the summer of 2009, a train of epic typhoons rolled across the Pacific. Ketsana dropped record rain on Manila and Vietnam; Morakot dumped nine and a half *feet* of rain on parts of Taiwan. All together? According to the *New York Times*, "the last thirty years have yielded four times as many weather-related disasters as the first three quarters of the 20th century combined."[31]

But lay aside hurricanes and wreckage. Just concentrate for a minute on how the sea is changing. For far longer than human civilization, those globe-girdling oceans have been chemically constant. They're so vast that we've taken their stability as a given. Even most oceanographers were shocked a few years ago when researchers began noticing that the seas were acidifying as they absorbed some of the carbon dioxide we've poured into the atmosphere. "It's been thought pH in the open oceans is well buffered, so it's surprising to see these fluctuations," said the University of Chicago biologist Timothy Wootton, who found acid levels rising ten times faster than expected.[32] Already ocean pH has slipped from 8.2 to 8.1; take one of those strips you dip in a hot tub, and you can tell the difference. The consensus estimate is that the pH will reach 7.8 by century's end.[33] The sea is already 30 percent more acid than it would have been because of our emissions, a process that Britain's Royal Society described as "essentially irreversible."[34] Already the ocean is more acid than anytime in the last eight hundred thousand years, and at current rates by 2050 it will be more corrosive than anytime in the past 20 million years. In that kind of environment, shellfish can't make thick enough shells. (Think of DDT and birds' eggs if you want an analogy.) By the summer of 2009, the Pacific oyster industry was reporting 80 percent mortality for oyster larvae, apparently because water rising from the ocean deep was "corrosive enough to kill the baby oysters."[35] At a conference in the spring of 2009, the American researcher Nancy Knowlton put it with refreshing bluntness: "Coral reefs will cease to exist as physical structures by 2100, perhaps 2050."[36] "We are overwhelming the system," says Richard Zeebe, an assistant professor of oceanography at the University of Hawaii. "It's pretty outrageous what we've done."[37] Which is as objective a scientific statement as you're likely to hear.

The idea that humans could fundamentally alter the planet is new. The Swedish chemist Svante Arrhenius broached the notion a century ago that we were "evaporating our coal mines into the air," and calculated that this would eventually raise temperatures, but nobody paid much attention. It wasn't until the 1950s that scientists even began measuring the amount of carbon dioxide in the atmosphere, from a small hut on the side of Hawaii's Mauna Loa, and they found that indeed the atmospheric concentration was steadily rising. But we didn't have the computing power to know what to make of that until the early 1980s, when a few research teams began investigating, and almost nobody outside of a few labs had heard of the notion until a NASA scientist named James Hansen testified before Congress in June 1988 that global warming was almost certainly beginning. Even then, though, the people most worried about the problem called it a future threat: the declaration that concluded the huge Rio summit on the environment in 1992 didn't even mention climate change, but did recommend, meekly, that "in order to protect the environment, the precautionary approach shall be widely applied by States according to their capabilities." People spoke mostly about global warming in the future tense; the word was always *threat,* right up through the 2008 presidential campaign. Unveiling his global warming initiatives at the University of New Hampshire, Barack Obama sounded a familiar note: "This is our generation's moment to save future generations from global catastrophe."

Here's his opponent, John McCain, a few months later: "We and the other nations of the world must get serious about substantially reducing greenhouse gas emissions in the coming years or we will hand off a much-diminished world to our grandchildren."[38]

In fact, if you've got a spare month some time, google *global warming* and *grandchildren.* Among the 585,000 essentially identical and anodyne responses:

Ted Kennedy, to Congress in 2008: "I cannot look into the eyes of my grandchildren and tell them: Sorry, I . . . can't do anything about it."

Barbara Boxer, at the National Press Club: "Will our grandchildren know the thrill of holding their child's hand watching with excitement a towering snow-capped mountain or awesome, calving glaciers?"

Arnold Schwarzenegger, signing new energy legislation: "I want to make California No. 1 in the fight against global warming.

This is something we owe our children and grandchildren." And Arnold at the United Nations: "We hold the future in our hands. Together we must ensure that our grandchildren will not have to ask why we failed to do the right thing, and let them suffer the consequences." And in a statement he e-mailed to the Chinese news agency Xinhua explaining the state's new mileage laws: "Last month I signed an Executive Order creating the world's first Low Carbon Fuel standard so our vehicles will emit less carbon and bring a healthier future to our children and grandchildren." Hasta la vista, grandchildren!

Joe Lieberman: "Shame on us if 100 or 200 years from now our grandchildren and great-grandchildren are living on a planet that has been irreparably damaged by global warming, and they ask, 'How could those who came before us . . . have let this happen?' "

David Attenborough: "If we do care about our grandchildren then we have to do something."

Former Illinois governor Rod Blagojevich, addressing his Climate Change Advisory Group: "By committing ourselves to action in Illinois, we can help minimize the effects of climate change and ensure our children and grandchildren inherit a healthy world full of opportunity."

The late Jerry Falwell: "I can tell you, our grandchildren will laugh at those who predicted global warming. We'll be in global cooling by then, if the Lord hasn't returned. I don't believe a moment of it. The whole thing is created to destroy America's free enterprise system and our economic stability."

Sir Richard Branson, chair of Virgin Airways: "I think businesses can influence leaders who are not worrying enough about our grandchildren."

Bill Clinton, stumping for his wife in Colorado: "We just have to slow down our economy and cut back our greenhouse gas emissions 'cause we have to save the planet for our grandchildren."

Let's let the movie critic Roger Ebert sum up the general feeling, in his review of Al Gore's *An Inconvenient Truth*: "You owe it to yourself to see this film. If you do not, and you have grandchildren, you should explain to them why you decided not to."

So how did it happen that the threat to our fairly far-off descendants, which required that we heed an alarm and adopt precautionary principles and begin to take measured action lest we have a crisis for future generations, et cetera—how did that suddenly turn into the Arctic melting away, the tropics expanding, the ocean turning acid? How did time dilate, and "100 or 200 years from now" become yesterday?

The answer, more or less, is that global warming is a huge experiment. We've never watched it happen before, so we didn't know how it would proceed. Here's what we knew twenty years ago: the historic level of carbon dioxide in the atmosphere, the level that produced those ten thousand years of stability, was roughly 275 parts per million. And also this: since the dawn of the Industrial Revolution we'd been steadily increasing that total, currently raising it more than two parts per million annually. But no one really knew where the red line was—it was impossible to really know in advance at what point you'd cross a tripwire and set off a bomb. Like, say, melting all the ice in the Arctic.

The number that people tossed around for about a decade was 550 parts per million. Not because we had any real data showing it was the danger point, but because it was double the historic concentration, which made it relatively easy to model with the relatively crude computer programs scientists were using. One paper after another predicted what would happen to sea levels or forest composition or penguin reproduction if carbon dioxide levels doubled to 550 parts per million. And so—inevitably and insidiously—that's the number we fixated on. Since it wouldn't be reached until the middle of the twenty-first century, it seemed to offer a little margin; it meshed plausibly with political time, with the kind of gradual solutions leaders like to imagine. That is, a doubling of carbon dioxide would happen well beyond the time that anyone now in power was likely to still be in office, or still running the company. It was when everyone's *grandchildren* would be in charge. As late as 2004, the journalist Paul Roberts, in his superb book *The End of Oil*, was able to write quite correctly that "most climate models indicate that once concentrations exceed 550 ppm we will start to witness 'dangerous' levels of warming and damage, especially in vulnerable areas, such as low-lying countries or those already suffering drought." But by then some doubt was beginning to creep in. Odd phenomena (large chunks of the Antarctic falling into the ocean, say) were unnerving scientists enough that, in Roberts's words, most "would much rather see concentrations stabilized at 450 ppm . . . where we might avoid most long-term effects and instead suffer a kind of 'warming light,' moderate loss of shorefront land, moderate loss of species, moderate desertification," and so on. And since even 450 was still 15 percent above our current levels, "we have a little room to breathe, which is handy."[39]

Or would have been. But as it turns out, we had been like commentators trying to call an election on the basis of the first precinct to report. Right about 2005 the real returns began to flood in, *flood* being the correct verb. And what they showed was that those old benchmarks—550, 450—had been wishful thinking. No breathing room, not when hurricane seasons like 2005 were setting new records for insurance payouts, not when polar ice was melting "fifty years ahead of schedule," not when the tropics "appear to have already expanded during only the last few decades of the 20th century by at least the same margins as models predict for this century."[40] Indeed, "ahead of schedule" became a kind of tic for headline writers: "Arctic Melt-off Ahead of Schedule" (the *Christian Science Monitor*, which quoted one scientist as saying "we're a hundred years ahead of schedule" in thawing Greenland), "Dry Future Well Ahead of Schedule" (the *Australian*), "Acidifed Seawater Showing Up Along Coast Ahead of Schedule" (the *Seattle Times*). The implication was that global warming hadn't read the invitation correctly and was showing up at four for the reception instead of six. In fact, of course, the "schedule" was wrong. And of course it was wrong—this was, as I've said, a huge experiment. Twenty-five years ago almost nobody even knew the planet was going to warm at all, never mind how fast.

It was that summer melt of Arctic ice in 2007 that seemed to break the spell, to start raising the stakes. The record minimums for ice were reached in the last week of September; in mid-December James Hansen, still the planet's leading climatologist, gave a short talk with six or seven slides at the American Geophysical Union meeting in San Francisco. What he said went unreported at the time, but it may turn out to be among the most crucial lectures in scientific history. He summarized both the real-world data that had emerged in recent years, including the ice-melt, and also the large body of research on paleoclimate—basically, the attempt to understand what had happened in the distant past when carbon dioxide levels climbed and fell. Taken together, he said, these two lines of inquiry made it clear that the safe number was, at most, 350 parts per million.

The day Jim Hansen announced that number was the day I knew we'd never again inhabit the planet I'd been born on, or anything close to it. Because we're already past 350—way past it. The planet has nearly 390 parts per million carbon dioxide in the atmosphere. We're too high. Forget the grandkids; it turns out this was a problem for our *parents*.

We can, if we're very lucky and very committed, eventually get the number back down below 350. This book will explore some of the reasons this task will be extremely hard, and some of the ways we can try. The planet can, slowly, soak up excess carbon dioxide if we stop pouring more in. That fight is what I spend my life on now, because it's still possible we can avert the very worst catastrophes. But even so, great damage will have been done along the way, on land and in the sea. In September 2009 the lead article in the journal *Nature* said that above 350 we "threaten the ecological life-support systems that have developed in the late Quaternary environment, and severely challenge the viability of contemporary human societies."[41] A month later, the journal *Science* offered new evidence of what the earth was like 20 million years ago, the last time we had carbon levels this high: sea levels rose one hundred feet or more, and temperatures rose as much as ten degrees.[42] The Zoological Society of London reported in July 2009 that "360 is now known to be the level at which coral reefs cease to be viable in the long run."[43]

We're not, in other words, going to get back the planet we used to have, the one on which our civilization developed. We're like the guy who ate steak for dinner every night and let his cholesterol top 300 and had the heart attack. Now he dines on Lipitor and walks on the treadmill, but half his heart is dead tissue. We're like the guy who smoked for forty years and then he had a stroke. He doesn't smoke anymore, but the left side of his body doesn't work either.

Consider: On January 26, 2009, less than a week after taking office, Barack Obama announced a series of stunning steps designed to dramatically raise fuel efficiency for cars. He also named a new envoy to aggressively negotiate an international accord on global warming. "This should prompt cheers from California to Maine," the head of one environmental group exulted. "The days of Washington dragging its heels are over," insisted the president.[44] It was the most auspicious day of environmental news in the twenty years of the global warming era. And then that afternoon, the National Oceanic and Atmospheric Administration released a new study showing that a new understanding of ocean physics proved that "changes in surface temperature, rainfall, and sea level are largely irreversible for more than a thousand years after carbon dioxide emissions are completely stopped." Its author, Susan Solomon, was interviewed on National Public Radio that night. "People have imagined that if we stopped emitting carbon dioxide that the climate would be back to normal in one hundred years or two hundred years," she said. "What we're showing here is that that's not right."[45] No one is going to refreeze the Arctic for us, or restore the pH of the oceans, and given the momentum of global warming we're likely to cross many more thresholds even if we all convert to solar power and bicycles this afternoon.

Which, it must be said, we're not doing. The scientists didn't merely underestimate how fast the Arctic would melt; they overestimated how fast our hearts would melt. The Intergovernmental Panel on Climate Change, or IPCC, carefully calculated a variety of different "emissions pathways" for the future, ranging from a world where we did everything possible to make ourselves lean and efficient to a "business-as-usual" model where we did next to nothing. In the last decade, as the United States has done very little to change its energy habits, and as the large Asian economies have come online, carbon emissions have risen "far above even the bleak scenarios" considered in the reports. In the summer of 2008, at an academic conference at Britain's Exeter University, a scientist named Kevin Anderson took the podium for a major address. He showed slide after slide, graph after graph, "representing the fumes that belch from chimneys, exhausts and jet engines, that should have bent in a rapid curve towards the ground, were heading for the ceiling instead." His conclusion: it was "improbable" that we'd be able to stop short of 650 parts per million, even if rich countries adopted "draconian emissions reductions within a decade." That number, should it come to pass, would mean that global average temperatures would increase something like seven degrees Fahrenheit, compared to the degree and a half they've gone up already.

"As an academic I wanted to be told it was a very good piece of work and that the conclusions were sound," Anderson said. "But as a human being, I desperately wanted someone to point out a mistake, and to tell me we had got it completely wrong." According to David Adam's account in the *Guardian*, nobody did. "The cream of the UK climate science community sat in stunned silence." In fact, Adam conducted a small poll himself among researchers, politicians, and activists. "Ask for projections around the dinner table after a few bottles of wine, and more vote for 650 ppm than 450 ppm as the more likely outcome," he reported.[46] Though the economic downturn that took hold in 2009 has as least temporarily slowed the rise—in fact American carbon dioxide emissions were expected to fall nearly 5 percent in 2009.[47] Which is good news. Just not good enough. To give you an idea of how aggressively the world's governments are willing to move, in July 2009 the thirteen largest emitters met in Washington to agree on an "aspirational" goal of 50 percent cuts in carbon by 2050, which falls pretty close to the category of "don't bother."[48]

The Copenhagen conference, in December 2009, was supposed to be the place where the world took an "historic step forward." Instead, it turned into a fiasco of the first order. Sure, there were giant rock concerts and a spirited protest march and twenty thousand environmentalists from around the world who showed up to lobby the talks. And there was actually powerful resistance to a meaningless deal from most of the nations of the world—the poor countries and the low-lying island nations stuck hard to their assessment that without deep cuts in emissions from the rich countries their very survival was at stake. Well more than half the nations of the world endorsed a strong target of 350 parts per million; the great cathedral in the center of the city, and then thousands of the world's other churches, rang their bells 350 times on the Sunday in the middle of the negotiations.

But the very next day the UN started locking the nongovernmental organizations out of the conference. An internal paper, leaked to the world's press, showed that even the UN knew the whole process was half-sham, because the proposed deals would increase temperatures much faster than the official rhetoric described. (My name was scrawled across the front, but I didn't leak it.) At week's end President Obama jetted in to "show leadership" and "break the deadlock," but all he did was repeat America's standing offer—by 2020 we'll cut our carbon emissions 4 percent below 1990 levels, a pledge whose stunning weakness his aides continued to blame on the difficulty of getting anything tougher through Congress. Fearing a face-destroying collapse, Obama negotiated a brief "Copenhagen accord" with the Chinese that lacked any targets or time frame for emissions, and then the president jetted out of town, eager to beat a snowstorm descending on Washington. The next day virtually every newspaper in the world declared it a debacle. As Joss Garman put it in London's *Independent*: "It is no exaggeration to describe the outcome of Copenhagen as a historic failure that will live in infamy."

But as usual you didn't need words to make the point at all, because numbers would do. A team of computer jockeys from MIT and elsewhere formed a group called Climate Interactive and built, in the months before Copenhagen, a sophisticated software model that could instantly analyze any proposal and tell you what it would mean a hundred years down the road. Here's what they found: if you took every government pledge made during the conference and added it all together, the world in 2100 would have *more than 725 parts per million carbon dioxide*, or slightly double what scientists now believe is the maximum safe level of 350. Even if you took all the possible "conditional proposals, legislation under debate and unofficial government statements"—in other words, even if you erred on the side of insane optimism—the world in 2100 would have about 600 parts per million carbon dioxide. That is, we'd live if not in hell, then in some place with a very similar temperature.

So far we've been the cause for the sudden surge in greenhouse gases and hence global temperatures, but that's starting to change, as the heat we've caused has started to trigger a series of ominous feedback effects. Some are fairly easy to see: melt Arctic sea ice, and you replace a shiny white mirror that reflects most of the incoming rays of the sun back out to space with a dull blue ocean that absorbs most of those rays. Others are less obvious, and much larger: booby traps, hidden around the world, waiting for the atmosphere to heat.

For instance, there are immense quantities of methane—natural gas—locked up beneath the frozen tundra, and in icy "clathrates" beneath the sea. Methane, like carbon dioxide, is a heat-trapping gas; if it starts escaping into the atmosphere, it will add to the pace of warming. And that's what seems to be happening, well ahead (need it be said) of schedule. In 2007, atmospheric levels of methane began to spike. Scientists weren't sure where they were coming from, but the fear was that those tundra and ocean sources were starting to melt in earnest. In the summer of 2008, a Russian research ship, the *Jacob Smirnitskyi*, was cruising off the country's northern coast in the Laptev Sea when the scientists on board started finding areas of the water's surface foaming with methane gas. Concentrations were a hundred times normal. "Yesterday, for the first time, we documented a field where the release was so intense that the methane did not have to dissolve into the sea water but was rising as methane bubbles to the sea surface," one of the scientists e-mailed a journalist at the *Independent.* "These methane chimneys were documented on an echo sounder and with seismic instruments."[49] The head of the research team, Igor Semiletov of the University of Alaska in Fairbanks, noted that temperatures over eastern Siberia had increased by almost ten degrees in the last decade. That's melting permafrost on the land, and hence more relatively warm water is flowing down the region's rivers into the ocean, where it may in turn be melting the icy seal over the underwater methane. The melting permafrost is also releasing methane on land. "On helicopter flights over the delta of the Lena River, higher methane concentrations have been measured at altitudes as high as 1,800 meters," reported Natalia Shakhova, of the Russian Academy of Sciences.[50] In recent winters scientists have reported that far northern ponds and marshes stayed unfrozen even in the depths of winter because so much methane was bubbling out from underneath. "It looks like a soda can is open underneath the water," one researcher explained.[51]

That's scary. Scarier even than the carbon pouring out of our tailpipes, because we're not directly releasing that methane. We burned the coal and gas and oil, and released the first dose of carbon, and that raised the temperature enough to start the process in motion. We're responsible for it, but we can't shut it off. It's taken on a life of its own. One recent estimate: the permafrost traps 1,600 billion tons of carbon. A hundred billion tons could be released this century, mostly in the form of methane, which would have a warming effect equivalent to 270 years of carbon dioxide emissions at current levels. "It's a kind of slow-motion time bomb," said Ted Schuur of the University of Florida in March 2009. At a certain point, he added, "the feedback process would continue even if we cut our greenhouse emissions to zero."[52]

We don't know if methane release has begun in earnest yet, or the exact threshold we'd need to pass. But there are dozens of such feedback loops out there. Peat covers about 2 percent of the planet's land surface, mostly in the far north—think moors, bogs, mires, swamp forests. They are wet places filled with decaying vegetation, a kind of nursery for what in many millennia could become coal. Because they're wet, they're very stable; the plants decompose very very slowly, so peatlands make a perfect "sink" for carbon, holding perhaps half as much as the atmosphere. But say you raise the temperature and hence the rate of evaporation; the water table starts to fall, and those swamps start to dry out. And as they do, the carbon in all that decaying vegetation starts to decompose more quickly and flood into the atmosphere. A 2008 study found, in fact, that "peatlands will quickly respond to the expected warming in this century by losing labile soil organic carbon during dry periods." How much? Well, peat bogs worldwide hold the equivalent of sixty-five years of fossil-fuel burning, and the expected warming will dry out enough of them to cause the loss of between 40 and 86 percent of that carbon.[53] It's as if we'd conjured up out of nowhere a second human population that's capable of burning coal and oil and gas nearly as fast as we do.

At the same time that we're triggering new pulses of carbon into the atmosphere, we're also steadily weakening the natural systems that pull it out of the air. Normally—over all but the last two hundred years of human civilization—the carbon dioxide level in the atmosphere remained stable because trees and plants and plankton sucked it up about as fast as volcanoes produced it. But now we've turned our cars and factories into junior volcanoes, and so we're not just producing carbon faster than the plant world can absorb it; we're also making it so hot that the plants absorb less carbon than they used to. In a 2008 experiment, scientists carved out small plots of grassland and installed them in labs where they could heat them artificially. "During this anomalously warm year and the year that followed, the two plots sucked up two-thirds less carbon than the plots that had been exposed to normal temperatures," the researchers reported.[54] The same thing may be happening at sea, where in January 2009 scientists "issued a warning" after finding "a sudden and dramatic collapse in the amount of carbon emissions absorbed" in fast-warming areas of the Sea of Japan.[55] Imagine that you desperately need to bail out your boat, but you find that your buckets are filled with holes that keep getting larger. "Fifty years ago, for every ton of CO_2 emitted to the atmosphere, natural sinks removed 600 kilograms. Currently the sinks are removing only 559 kilograms per ton, and the amount is falling."[56] Those are big holes.

So far I've written more about causes than effects; before long we'll begin to see how these new realities play out, how they build on each other in a crescendo of cascading consequences. But here, toward the start, I want simply to establish the bottom line. We're changing the most basic dynamics of the only world we've ever known.

The only truly crucial question that human beings ask is: "What's for dinner?" Or, for much of human history, "Is there any dinner?" At an international meeting in Poland in December 2008, Martin Parry, one of the cochairs of the Intergovernmental Panel on Climate Change, gave a talk that began like this: "The 2008 food crisis is the largest impact of climate change so far. It was caused partly by the poorly-thought-through switch to biofuels as a way of combating climate change, and partly by the drought in western Australia, which local scientists have identified as having been caused by climate change." The result, Parry said, was that in 2008, 40 million people had been added to the list of those "at risk of hunger," taking the total to 963 million, or one-sixth of the world's population.[57] That is, in one year climate change had managed to turn 40 million people—more than the population of California—hungry. Not "this could happen." This happened. And in 2009 the number topped a billion.[58]

In January 2009, a team analyzing twenty-three climate models told us about the future. They compared the expected new temperatures by century's end with what we know about wheat and corn. They found that it will routinely get so hot that the crops will grow much less vigorously; wheat yields could easily fall 20 to 40 percent, on a planet that's expected to host 3 billion more people. We've already begun to see this in action. In 2003, France had the kind of heat wave that will become the new normal as the decades roll on. Not only did thirty thousand people die because of heat stress, but corn production fell by a third, fruit harvests by a quarter, and wheat by a fifth. The jovial notion that we'll compensate by simply moving farther north eventually becomes absurd. "You can't move that far north because all you end up with is pretty infertile tundra," one of the researchers, the University of Washington's David Battisti, pointed out. "When all the signs point in the same direction, and in this case it's a bad direction, you pretty much know what's going to happen," he said. "You are talking about hundreds of millions of additional people looking for food because they won't be able to find it where they find it now."[59] The chief scientist at the U.S. State Department said recently that its analysis foresees famines severe enough to affect a billion people at a time in the next few decades; Britain's chief scientist said in the spring of 2009 that "a perfect storm" of food and water shortages could hit by 2030.[60] Here's the Stanford University researcher Rosamond Naylor, who conducted some of the most recent calculations: "I think what startled me the most is that when we looked at our historic examples there were ways to address the problem within a given year. People could always turn somewhere else to find food. But in the future there's not going to be any place to turn."[61] It doesn't get any more basic than that.

Or maybe it does get more basic, since we're not the only species involved. Often, speaking to audiences, I'll find people who have moved to a zone of spooky calm: yes, they say, human beings may do themselves in, but "the planet" will survive. That's true in some sense, at least until the sun explodes, but it won't be anything like the planet we've known. We're hard at work transforming it—hard at work sabotaging its biology, draining its diversity, affecting every other kind of life that we were born onto this planet with. We're running Genesis backward, decreating. Melt the Arctic, for instance, and you wreak havoc with the region's phytoplankton, "the crucial nutrient at the base of the food web on which marine life depends."[62] In the far South, a 2008 study noted, three-fourths of big penguin colonies may soon disappear.[63] I've stood in the middle of these rookeries, a hundred thousand mating pairs shrieking, their babies demanding food. It's the greatest example of fecundity I've ever seen—you can smell them miles away. They define the insane abundance of the world we've known, and their absence will help to define the new world we're creating.

The changes could hardly be more fundamental. For instance, a team of scientists showed recently that all manner of animals are likely to shrink, literally, as temperatures climb. Larger animals have a lower ratio of surface area to volume, so they retain heat more easily and do better in cooler climes, whereas smaller species radiate heat more easily. "It makes sense to be bigger when it's colder," says Wendy Foden, a biologist at the World Conservation Union. "As the world gets warmer, species will shrink."[64] In July 2009, researchers in fact found that Scottish sheep had been shrinking three ounces a year for two decades because of warmer temperatures; the same with red-winged gulls and certain crustaceans. "Whether in the future we're going to get miniature bonsai sheep I have no idea," said a biologist at Imperial College in London.[65]

And as the world gets warmer, it also gets steadily simpler. "From Peru to Namibia to the Black Sea to Japan . . . massive swarms of jellyfish are blooming," researchers said in 2008, "closing beaches and wiping out fish, either by devouring their eggs and larvae, or out-competing them for food." In the Sea of Japan, 500 million Nomurai jellyfish—each more than two meters in diameter—are clogging fishing nets; a region of the Bering Sea is so full of jellies that it's been renamed "Slime Bank." "Jellyfish grow faster and produce more young in warmer waters," one researcher explained.[66] The fish and whales that remain live in a world changing as fast as ours, in every way. New studies show, for instance, that as seawater grows more acid, it absorbs less sound, making the whole ocean noisier. As one scientist put it. "It's the cocktail party effect."[67] Meanwhile, scientists reported in October 2009 that "as sea temperatures have risen in recent decades, enormous sheets of a mucus-like material have begun to form" in the world's seas. Some of these "blobs" are two hundred kilometers long, carry high levels of

E. coli bacteria, and often "trap animals, coating their gills and suffocating them."[68]

Here's all I'm trying to say: The planet on which our civilization evolved no longer exists. The stability that produced that civilization has vanished; epic changes have begun. (My favorite bleak headline, from *USA Today* in May 2009, describes a new study from the American Meteorological Society: "Global Warming May Be Twice as Bad as Previously Expected.")[69] We *may*, with commitment and luck, yet be able to maintain a planet that will sustain *some kind* of civilization, but it won't be the same planet, and hence it can't be the same civilization. The earth that we knew—the only earth that we ever knew—is gone.

If that stable earth allowed human *civilization*, however, something else created *modernity*, the world that most of us reading this book inhabit. That something was the sudden availability, beginning in the early eighteenth century, of cheap fossil fuel. An exaggeration? One barrel of oil yields as much energy as twenty-five thousand hours of human manual labor—more than a decade of human labor per barrel. The average American uses twenty-five barrels each year, which is like finding three hundred years of free labor annually. And that's just the oil; there's coal and gas, too.[70] It's why most of the people reading this book don't do much manual labor anymore, and why those who do use machines that make them hundreds of times more powerful than their forebears. It's why we're prosperous, why our economies have grown. It's also, of course, why we have global warming and acid oceans; in essence we've spent two hundred years digging up all that ancient carbon, combining it with oxygen for a moment to explode the pistons that take us to the drive-through, and then releasing it into the atmosphere, where it accumulates as carbon dioxide. That cloud of carbon is nothing more than a ghostly reflection of the pools of oil and veins of coal where it once dwelled—each gallon of gasoline represents a hundred tons of ancient plants.[71] All day every day we burn coal and gas and oil, from the second we make the coffee till the second we turn out the lights. (And is the furnace still running? The air-conditioning?) If an alien landed in the United States on some voyage of exploration, he might well report back to headquarters that we were bipedal devices for combusting fossil fuel.

Which is why it's unlucky in the extreme that at precisely the same moment that we've destabilized the climate that underwrote civilization, we've also started to come up short on the fossil fuel that underwrote modernity. The two phenomena (very much intertwined) have struck us with the same uncanny speed. Just as a few scientists began warning a generation ago about rising temperatures, so a tiny band of geologists began fretting about dwindling oil supplies. In 1956, two years before the first carbon dioxide monitor was installed on Mauna Loa, a petroleum geologist named M. King Hubbert first predicted that U.S. oil production would reach its zenith between 1965 and 1970. He was spot-on—but nobody worried too much, because so much oil was flowing in from the great fields of the Middle East. In recent years, however, there have been troubling signs that those fields, too, are starting to dwindle, and clear evidence that no new fields big enough to make up for their decline have been discovered. "Peak oil" began as a fringe idea—just like climate change—but in recent years more and more establishment figures have signed on to the idea that we may really be reaching the point where the amount of oil we can wrest from the planet will go down, not up.

The debate ended on November 12, 2008. If you didn't notice, blame post-Obama hangover or the ragged fear (and low oil prices) that came with the height of the financial crisis. November 12 was the day the Bush administration decided to stop buying up toxic assets and instead just recapitalize the banks, and the day that Obama named his transition team. But the real news that day, the data that rewrote the history books, came from the International Energy Agency, which published its long-awaited World Energy Outlook. The IEA defines conservative—it's the group set up by rich nations in the wake of the oil shocks of the 1970s to maintain a steady supply of energy. And their economists had always insisted that there would be a growing supply of oil for decades to come. No problem, no problem, no problem. Plenty of oil.

This time around, the tune changed markedly. First, said the IEA, production in current oil fields is falling by about 7 percent a year, a figure that will rise steadily to 9 percent over the next few decades. In other words, the level of oil in these giant fields has dropped far enough that we can no longer get as much as we used to. Never mind fueling the growing Asian thirst for oil; simply running in place would mean finding four new Saudi Arabias by 2030. But since demand *will* keep rising in Asia (92 percent of American adults own cars, compared with 6 percent of Chinese) and elsewhere, staying abreast will mean finding *six* new Saudi Arabias—or a new Kuwait—every year. The IEA put it in dollar figures: keeping up our oil economy will require $350 billion in exploration and investment every year through 2030. That's compared with a total of $390 billion that the world spent on those items in the whole period of 2000–2007, when the economy was booming.[72] And even the IEA's gloom may well have been too optimistic. A few weeks later, Merrill Lynch energy analysts, using new numbers for non-OPEC oil fields, calculated that we'll need *ten* new Saudi Arabias by 2030.[73] As the former

CIA director and defense secretary James Schlesinger put it, "The battle is over, the oil peakists have won."[74]

On the old planet—the one with an Arctic ice cap, the one where hurricanes didn't strike Spain and Brazil, the one where jellyfish didn't bloom in great slimy clouds across the oceans—we had one Saudi Arabia and one Kuwait. They sat atop enormous pools of oil. Now, every day more so, they sit atop big empty holes. And there are no more Saudi Arabias, no matter how much money you have. So does modernity disappear along with the oil? It's a question worth asking, when six of the twelve largest companies in the world are fossil-fuel providers, four make cars and trucks, and one, General Electric, is, as its name implies, heavily involved in the energy industry. Just buying fossil fuel requires almost a tenth of global GDP, and almost all the other 90 percent depends on burning the stuff.[75]

Oil is also the mother of most petrochemicals and plastics. Richard Heinberg, the analyst who was one of the first to alert the world to the impending oil peak, once compiled a list of things made from oil that ran from computer chips, insecticides, anesthetics, and fertilizers, right through lipstick, perfume, and pantyhose, to aspirin and parachutes. "Without petrochemicals," Heinberg wrote, "medical science, information technology, modern cityscapes, and countless other aspects of our modern technology-intensive lifestyles would simply not exist. In all, oil represents the essence of modern life."[76] That we've wasted it so mindlessly is depressing. (From the mid-1980s on American automakers stopped worrying about efficiency and instead concentrated on torque; as a result, by 2002 the average American car would go from zero to sixty in 10.5 seconds, a dynamic 3.5 seconds faster than a generation earlier.)[77] But it's also understandable. Again: cheap energy is not a useful part of our economy. It *is* our economy. "Before 1850 most Americans didn't even know coal could be burned," writes Paul Roberts. "Yet by 1900 U.S. mines were outproducing those in England. What were people using all this extra energy for? Mainly people were manufacturing more things: more textiles, more machines, more food and ale, more paper. The pattern was clear: the more you produced, the more energy you needed. And conversely, the more energy you used, the more things you produced."[78]

Because there was lots of it on that old planet, energy was cheap. You've seen the pictures—the early oil strikes where the fields were under such high pressure that as soon as you punctured them with a drill the crude would spew into the air. It was, more or less, free for the taking. No more; what's left is in hard-to-get-at places and requires fantastic technical skill. Norway's Troll A platform in the North Sea, for instance, is the largest man-made structure ever moved: each of its three concrete legs is 994 feet long with an elevator that takes nine minutes to travel from the seabed to the drilling platform above. (To celebrate its tenth anniversary, a Norwegian pop idol sang a concert at the bottom of the elevator shaft, the deepest musical performance in history.) All of which means that drilling oil is getting progressively more expensive, not just in dollar terms but, more important, in what economists call "energy return on investment," or EROI. If the EROI on an oil well is 20:1, you get twenty units of energy out for every unit you put in. Twenty to one is pretty good—a lot better than, say, taking Canadian tar sands and melting them down to get usable oil. That might produce an EROI of 5.2:1 by some recent estimates. Corn ethanol for oil? Once you've figured in all the energy it takes to grow the stuff and process it, you're lucky to break even.[79] Charles Hall, a professor at the State University of New York, argued recently that "to offer any remotely viable contribution to society, a liquid fuel should not be dependent on subsidies from petroleum and should have an EROI of at least 5:1." Solar panels:

somewhere between 2.5 and 4.3:1, at least for now.[80] Which is not to say that solar panels are a bad idea—this book is being written with juice flowing straight from my roof. Only that they won't replace fossil fuels straight up.

We got a taste of that in the remarkable spring of 2008, as oil prices started to rise through the roof. Economies were strong, demand was rising—and there was no new supply to meet it. Paul Roberts pointed out in 2004 that six of the last seven global recessions had been preceded by spikes in the price of oil, and now we can safely make that seven of the last eight. Economic historians will long debate exactly why the economy keeled over in the fall of the year, but collapsing home prices seem to be the most basic answer. And they collapsed not just because of mortgage fraud but also because people began to take note of reality: in a world where four-dollar-a-gallon gas was even a possibility, who wanted a starter castle ninety minutes from work? "As oil prices started to bite, the new housing built in distant suburbs and even more remote 'exurbs' became less viable for commuters," wrote the oil analyst Phil Hart.[81] Between 2004 and 2008, when gas prices rose past two dollars to their eventual peak, the three cities with the largest declines in housing prices were the entirely auto-dependent Las Vegas, Phoenix, and Detroit; Portland, Oregon, the bike-and-trolley capital of the country, saw the largest rise in home value.[82]

But it's not only transportation. Since oil is in everything, its price affects the entire economy. In the spring of 2009, a University of California economist reported that "nearly all of last year's economic downturn could be attributed to the oil price shock"; despite his data, he reported, "it was a conclusion he didn't quite believe in himself," except that each of the previous run-ups in oil prices—1973, 1979, 1990, even 2001—also corresponded with recessions.[83] Once the economy collapsed, of course, oil prices collapsed with them; we went back to consuming a little less than the planet was capable of producing. But should the economy recover, oil prices will almost certainly bounce right back. As the financier George Soros, who made a pile betting on the rise and fall of oil in 2008, wrote that autumn, "any relief will be temporary."[84]

In fact, one all-too-likely result of peak oil will be even more use of our most abundant fossil fuel, good old coal. And the certain result of using more coal will be . . . more global warming, since it's the dirtiest of all the fossil fuels, producing twice the carbon dioxide of oil. As James Hansen and his NASA team pointed out, any increased reliance on coal is enough to guarantee that we'll never get back to 350. Cue doom.

These are the kinds of traps we fall into on this new planet. We can't burn more oil because it's running out. The stuff we can still find to burn triggers even more global warming. The most vicious of cycles.

We know, definitively, that the old planet "worked." That is, it produced and sustained a modern civilization. We don't know that about the new one.

The traditional way of imagining the effects of climate change is simply to list disparate data points—to go around the world inventorying the items that one scientist or another has managed to model and predict. In a sense, to list the symptoms. So:

- Engineers in Dublin are convinced that higher tides caused by climate change are eroding the famous O'Connell Bridge that spans the River Liffey at the foot of the Irish capital's main thoroughfare.[85]
- A state of emergency was declared in the Marshall Islands late on Christmas Eve in 2008, as widespread flooding displaced

hundreds of islanders, the third time in two weeks that powerful storm surges had swamped the main cities of Majuro and Ebeye, each of which sits less than three feet above sea level. The floodwaters not only damaged houses and roads but also destroyed cemeteries.[86]

- "Tick drags" across my home state of Vermont are finding these agents of Lyme disease alive in the forest even in January and February. In the spring of 2008, the state entomologist Jon Turmel found thirty to forty ticks on his pant leg after walking twenty feet along the Connecticut river valley in the village of St. Johnsbury. He described the tick population in the area as "extreme." Indeed.[87]
- The residents of Ocean Isle Beach, North Carolina, are spending as much as thirty thousand dollars each to place giant sandbags in front of their homes in an effort to ward off the ocean. "There used to be a street in front of our house, and then a row of cottages," says Lisa Schaeffer. After Tropical Storm Hanna her home stood just five yards from the sea.[88]
- Along the Yukon River in Canada, warmer water has made Chinook salmon "more susceptible to the parasite *Ichthyophonus*. Subsistence farmers must now catch 150 salmon to yield 100 usable ones," according to a Natural Resource Defense Council study.[89]
- Reduced winter ice cover means that evaporation will proceed year-round, and hence the water level in Lake Erie could fall between three and six feet in the next seventy years, making shipping difficult (for every inch the lake drops, a commercial ship must leave behind 270 tons of cargo) and shifting the shoreline several miles in Sandusky Bay.[90] Moreover, the range of the official Ohio state symbol, the buckeye tree, may shift north, out of the state entirely and into the territory of its college football archrival, Michigan.[91]

- A Harvard study found that ragweed grows 10 percent taller and produces 60 percent more pollen as the temperature warms.[92]

The other time-honored method for communicating this kind of news is to find individual victims and share their stories, in the hope that narrative will accomplish what statistics can't. We don't pay much attention to poor people, so it can astonish us to read stories of just how hard life has become, like the ones John Vidal collected for London's *Guardian* in the fall of 2008.

- "Juan Antonio's eyes are full of tears," Vidal reports. "If good rains do not come, he says, he will pack his bag, kiss his wife and two children goodbye, and join the annual exodus of young men leaving hot, dry, rural northeast Brazil for the biofuel fields in the south." Droughts in the region are longer and more frequent now than in the past. "Climate change is biting," a Brazilian agronomist named Lindon Carlos tells him. "It is much hotter than it used to be and it stays hotter for longer."
- "It's far warmer now," says one Bangladeshi villager in the Deara district, whose only name is Selina. "We do not feel cold in the rainy season. We used to need blankets but now we don't. There is extreme uncertainty of weather. It makes it very hard to farm and we cannot plan. The storms are increasing and the tides now come right up to our houses."
- "Tekmadur Majsi farms in the upland Nepali village of Ketbari," Vidal writes. "Small floods once a decade or so are routine, but now they've grown larger and more common." Majsi is not hopeful for the future. "We always used to have a little rain each month, but now when there is rain it's very different. It's more concentrated and intense," he tells the reporter. "It means crop yields are going down."[93]

Vidal's reporting is not unique. Eliza Barclay of the *Miami Herald* traveled to the Cordillera Blanca, eleven thousand feet up in the Peruvian Andes, where she met a man named Gregorio Huanuco, who farmed as his ancestors had for generations. In 1990 Huanuco began to notice change: "a battering hailstorm, two months without rain, a warm winter. Then the quirky weather became more consistent and other oddities began to appear: rats nibbling away at his cereal crops and a fungus blanketing his potatoes." Huanuco's way of life was slipping away. "Before we planted all year long, any month we wanted to," he said. "Now we only get water a few times a year and so we cannot plant as much, and the pests and diseases keep coming."[94]

Or consider what Ben Simon, a reporter for Agence France-Presse, found on the slopes of Mount Speke, one of Uganda's highest peaks. The snowcap was almost gone, and farmers trying to eke out a living have to climb farther up the hill each year to find a climate cool enough to grow their beans. He quoted Nelson Bikalnumuli: "People just keep moving up, up, up. I fear soon we may be on top of each other."[95]

In Haiti, where an unprecedented four fierce hurricanes hit in quick succession in 2008, Marc Lacey of the *New York Times* found a mother living with her six children on a roof in the city of Gonaïves. "At the main cathedral, the water rushed in the front door, toppling pews and leaving the place stained with mud and smelling of sewage," he reported. "Upstairs, dozens of people have taken refuge, huddled together on the concrete floor. When a visitor arrived, they rubbed their bellies and pleaded for nourishment."[96]

Or perhaps you have a hard time identifying with poor peasants stranded in impoverished villages. Consider, then, the story of the MV *Nautica*, a stately liner of the Oceania Cruise company, in whose thousand-square-foot suites "every inch is devoted to your pleasure," with "euro-top mattresses," forty-two-inch plasma screens, wraparound teak verandas, and "a second bathroom for guests." (The ship's spa offers an "exotic lime and ginger salt glow with massage," or, more worryingly, an "exotic coconut rub and milk ritual.") Anyway, the *Nautica* set off on its thirty-two-day "Odyssey to Asia" in the summer of 2008 but had to scrap "three magical days in the capital of the former British colony of Burma," after Cyclone Nargis wrecked both the nation and its image. "Considering the destruction, we said, no, not a wise move to be scheduling a call there," one mate explained. But the compensating longer stay in Mumbai was scrapped, too, after terrorist attacks, and then cruising through the Gulf of Aden the ship was attacked by pirates who fired eight shots. "We didn't think they would be cheeky enough to attack a cruise ship," said Wendy Armitage of Wellington, New Zealand. So the *Nautica* reset course for the Maldives, where nothing bad happened.[97] Although the same month the liner was in port, the president of the Maldives announced that his low-lying nation was planning to save a billion dollars annually from its tourist income so that it could buy land and relocate the population to Sri Lanka or Australia before the ocean finally rose too high for its survival. "We will invest in land," he said to CNN. "We do not want to end up in refugee tents if the worst happens."[98] The Maldives weren't alone, by the way. A few months later the Pacific island nation of Kiribati announced a similar plan.[99]

The trouble with this endless collection of anecdotes, though, is that it misses the essential flavor of the new world we're constructing. Every individual problem, even if it's impossible to endure, is fairly simple and straightforward. The temperature rises, and the buckeye tree migrates north. The temperature rises, and the hurricanes get more frequent and you starve. The temperature rises, and the level of the ocean comes up and it floods your cemetery and you really can't live on your island anymore. The temperature rises, and even in the Mandara Spa on the Salon Deck, it's hard to imagine that "the nimble fingers of an able masseuse will soothe away all the cares of the world."[100] Simple, understandable.

In truth, though, our new planet is much more complex and interesting. It's not just that the things we used to do are getting harder; it's that these initial and obvious effects lead us into a series of double and triple binds that make *any* action hard. We don't really know where to turn, because the planet we now inhabit doesn't work the way the old one did. Sometimes the irony couldn't be clearer. We've already seen that the far North is melting fast. As the sea ice goes, the albedo, or reflectivity, of the Arctic changes, with the mirror of white ice replaced by sun-absorbing blue. And the permafrost melts, and the methane escapes, and peat bogs dry out and add to the load of carbon. But something else happens, too. All of a sudden you can start drilling for gas and oil in these places. The Arctic, by some estimates, may hold 20 percent of the planet's undiscovered reserves, not enough to hold off peak oil for very long but enough to guarantee one more pulse of carbon into the atmosphere.

Now try a slightly more complicated problem. We've been burning down rain forests for a long time to create cheap agricultural land in the Amazon, and that obviously puts carbon into the atmosphere. It was enough of a worry—remember all those "save the rain forest" concerts in the 1990s?—that Brazil started enforcing its conservation laws, and the rate of loss began to ebb. But as those holes grew beneath the Middle East, and oil became more expensive, the market for biofuels strengthened. All of a sudden soybean farmers started pushing deeper into the jungle; deforestation jumped 64 percent in 2008 as oil prices rose.[101] One observer reported watching "bulldozers operating like Panzer divisions leveling and burning forests."[102] Meanwhile, Britain's Meteorological Office released new research in November 2008 (the same week, in fact, as the IEA report on declining oil supplies), which showed that climate change was producing drier conditions over much of the region, making the rain forest more prone than ever to natural fires—within a decade much of southeast Amazonia would be in the zone of higher fire risk.[103] Those fires produce even more carbon, and by destroying the forest they also remove a natural sink for carbon. What is left behind is a hotter, drier clearing: African research shows that the daytime temperature in the soil above a cleared patch is eight degrees higher than in the nearby forest, and the humidity is 49 percent, compared with 87 percent in the forest.

Something like that appears to be what's happening across the tropics. In the Amazon, reports the researcher Peter Bunyard, "already we are seeing parts of the Basin drying out and forming savanna, with its drought-tolerant shrubs and grasses, in what may well be the beginnings of a savannizing process that could lead to desertification."[104] In normal times—that is, on the old earth—the Amazon managed to move water much farther inland from the oceans than the rain would normally fall. The first swath of jungle gets wet, and then transpires the moisture through its leaves, forming new clouds that produce new rainfall farther west—all in all, a series of six pulses that move the ocean's bounty all the way to the Andes. The energy involved is prodigious—the equivalent of 4 million or more atomic bombs' worth a day. The forest, in essence, is "a gigantic irreplaceable water pump," in Bunyard's phrase, which in turn powers much of the planet's current air circulation system, taking "energy out and away from the Amazon basin to the higher latitudes, to the more temperate parts of the planet. Argentina, thousands of miles away from the Amazon Basin, gets no less than half its rain courtesy of the rainforest, a fact that few, if any, of the Argentinian landowners are aware of. And in equal ignorance the U.S. receives its share of the bounty, particularly over the Midwest." In fact, studies show that rainfall over the Amazon Basin is paralleled, four months later, by spring and summer rain across the U.S. corn belt.[105]

All of this is wildly complicated. It is perhaps enough to say that the Amazon is one of our planet's largest physical features, and it is far more vulnerable than we'd assumed, both to the onslaught of deforestation for food and biofuels, and to the changes in temperature that we've kicked off. The net result of the various forces, Bunyard says, will be a "much-diminished rainfall regime over the Amazon," with "rapid forest dieback and death." Oh, and as that happens, the decomposition of all the old forest "may well lead to more than 70 gigatons of carbon escaping as carbon dioxide into the atmosphere."[106] Instead of the "lungs of the planet" sucking in carbon and breathing out oxygen, the great green jungle turns into one more smokestack.

But the Amazon is far away, mysterious. You've more likely been to the high forests of the North American West, to the Rockies and the Sierras—probably driven the Road to the Sun at Glacier Park, or motored over Donner Pass. Certainly you've looked at Ansel Adams's photographs—this is our iconic idea of the wild. These ranges are also, like the poles or the Amazon, key natural features on which we depend. As the *Sacramento Bee* once described it, the Sierra is "a giant water faucet in the sky, a 400-mile-long, 60-mile-wide reservoir held in cold storage that supplies California with more than 60 percent of its water, much of it when it's needed most: over the hot, dry summer months."[107] Already that snowpack has shrunk by more than 10 percent, with the forecast that it will shrink as much as 40 percent more by midcentury and as much as 90 percent by century's end.

But let's not speculate; let's just focus on what has already happened: "Temperatures have warmed during winter and early-spring storms," noted one study. "Consequently the fraction of precipitation that fell as snow declined, while the fraction that fell as rain increased." And when rain falls in the winter in the Sierras, bad things happen—the massive New Year's Day flood in 1997, for example, when rain fell as high up the mountains as eleven thousand feet and the ensuing deluge resulted in disaster declarations for all forty-six counties in northern California. California's four wettest winters on record have come since 1996; in 2008 the state's energy planners started conducting drills for dealing with epic floods that forecasters say are becoming ever more likely.[108] Something else happens when the snowpack melts early—the sun now has time to dry out the forest, guaranteeing a longer fire season and drier trees.[109] In fact, the average California fire season runs seventy-eight days longer than it did in the 1970s and 1980s; it used to start in June and end in September, but now the Forest Service hires firefighting crews in the middle of April, and they are often still working into November and December. Half the National Forest Service budget is now spent extinguishing fires: "The agency is no longer the U.S. Forest service but rather the U.S. Fire Service," one congressman complained.[110]

As with hurricanes, it's not just more fires but bigger ones. On average, large fires now burn four times as long as a generation ago, and in recent years three-quarters of the bad fires across the West came in years when the snow melted well ahead of schedule. "We're getting in a place where we are almost having a perfect storm" for wildfire, said one Forest Service official. And, of course, it all feeds back on itself. The Moonlight fire, in September 2007 near Lake Tahoe, burned for two weeks and in that time pumped an estimated 5 million tons of carbon dioxide into the atmosphere, the same as 970,000 cars driving for a year, the same impact as a coal-fired power plant. "The intensity of the fire was pretty spectacular," the incident commander told Tom Knudson of the *Sacramento Bee.* When it was over, even the soil was incinerated, making it hard for the conifer forest to return. Researchers now believe that more large fires will lead to thinner, scrubbier woods, and indeed, black oak, whitethorn manzanita, and other brush species are rapidly expanding across parts of the Sierra that once grew mostly pine. One result? Western forests, which are currently responsible for 20 to 40 percent of total U.S. carbon sequestration, may soon become a source of carbon dioxide, not a sink for the gas.[111] Another, just as depressing: the biggest trees, the largest living things on earth, are disappearing. A Yosemite study found in 2009 that the "density of large-diameter trees in the forest" has fallen by a quarter in recent decades. "These large, old trees have lived centuries and experience many dry and wet periods," one researcher said. "So it is quite a surprise that recent conditions are such that these long-term survivors have been affected." The decline could "accelerate" as the climate warms, the study adds.[112]

Let's move a few hundred miles east, to the spine of the Rockies, where trees are dying in incredible numbers. Partly it's chronic; heat stress and lack of water have doubled the "background mortality" of trees in the area.[113] But there's also acute trouble. By 2008 Wyoming and Colorado alone housed more than three million acres of dead trees.[114] In the next five years,

Colorado expects to lose another 5 million acres—virtually every lodgepole pine larger than five inches in diameter. Farther north, in British Columbia, 33 million acres of lodgepole have already turned from green to rust-red, all dead. The culprit is the mountain pine beetle, Latin name *Dendroctonus,* which translates as "tree killer." Once the beetle drills into the bark, the tree gives off a white, waxy resin in an attempt to seal the insect in its hole. But the attacker can give off a pheromone that draws swarms of other beetles. Eventually the tree is overwhelmed.[115] "The scope and scale of the destruction is like nothing we have ever seen," says Jay Jensen, executive director of the Council of Western State Foresters. "We're seeing the end of some forests as we know them."[116]

Why is it happening? Because we've raised the temperature enough that the beetles can overwinter more easily. Milder winters since 1994 have reduced the winter death rate of beetle larvae in Wyoming from 80 percent per year to less than 10 percent.[117] You need stretches of thirty or forty degrees below zero up in the mountains to kill off the beetles, and that doesn't happen much anymore. (In Glacier National Park, for instance, only 25 of the 150 glaciers that were there in 1850 still exist, and all of them are shrinking rapidly.)[118] Meanwhile, hotter, drier summers have made trees weaker and less able to fight off the swarming beetles. And what is the result? All the obvious things: greatly increased fire risk, followed by mudslide and erosion. Dead trees falling on roads and toppling power lines. In Colorado and Wyoming, officials closed thirty-eight campgrounds so trees wouldn't drop on tents. And a kind of despair. "It's really something to see," a Utah state forester said. "You would be very surprised. It's hard to describe until you see it—it's just dead trees as far as the eye can see."[119]

Oh, and this you'd never guess: lots more carbon flooding into the atmosphere. A study in the journal *Nature* in the fall of 2008 offered this tally: during outbreaks of pine beetle infestation, "the resulting widespread tree mortality reduces forest carbon uptake and increases future emissions from the decay of killed trees." Since these outbreaks are "an order of magnitude larger in area and severity than all previous recorded outbreaks," the impact "converted the forest from a small net carbon sink to a large net carbon source."[120] Indeed, in early 2009 the Canadian government, which had long argued that its carbon-sequestering forests should count against its tar-sand burning in UN tallies of its carbon dioxide output, quietly dropped the claim. Now that the trees have died, timber companies want to log them off, but environmentalists have pointed out that that would in turn release much of the carbon stored in the peaty soils beneath the trees, igniting what one called a "carbon bomb." By some estimates, Canada's forests alone contain 186 billion tons of carbon, or the equivalent of twenty-seven years of global emissions from burning coal and gas and oil.[121]

Once trends like this get rolling, we can't slow them. We don't know how to refreeze the Arctic or regrow a rain forest. Here's what it looks like: in the last six years, as warming temperatures and drought have killed off the native vegetation that holds soil in place, windstorms have dumped twice as much dust across the American West.[122] In April 2009, after the biggest of the storms blew through Silverton, Colorado, one witness said the landscape "looked like Mars. . . . You could feel the dust, you could taste the dust." But as usual the damage reverberates. The storms drop huge quantities of dirt on the snowpack of the Rocky Mountains, darkening the white ice and significantly speeding up its melt. "It's effectively like turning the sun up fifty percent," explains one University of Utah professor.[123] The snowpack now melts "weeks earlier than normal," according to Scott Streator of Greenwire, which spells "disaster for thousands of farmers and ranchers in the region who depend on slowly melting snow to provide water" flows over the dry summer months.[124] "A lot of the water's gone by the time the crops need it," one researcher explained.[125]

So let's review. The planet we inhabit has a finite number of huge physical features. Virtually all of them seem to be changing rapidly: the Arctic ice cap is melting, and the great glacier above Greenland is thinning, both with disconcerting and unexpected speed. The oceans, which cover three-fourths of the earth's surface, are distinctly more acid and their level is rising; they are also warmer, which means the greatest storms on our planet, hurricanes and cyclones, have become more powerful. The vast inland glaciers in the Andes and Himalayas, and the giant snowpack of the American West, are melting very fast, and within decades the supply of water to the billions of people living downstream may dwindle. The great rain forest of the Amazon is drying on its margins and threatened at its core. The great boreal forest of North America is dying in a matter of years. The great storehouses of oil beneath the earth's crust are now more empty than full. Every one of these things is completely unprecedented in the ten thousand years of human civilization. And some places with civilizations that date back thousand of years—the Maldives in the Indian Ocean, Kiribati in the Pacific, and many other island nations—are actively preparing to lower their flags and evacuate their territory. The cedars of Lebanon—you can read about them in the Bible—are now listed as "heavily threatened" by climate change.[126] We have traveled to a new planet, propelled on a burst of carbon dioxide. That new planet, as is often the case in science fiction, looks more or less like our own but clearly isn't. I know that I'm repeating myself. I'm repeating myself on purpose. This is the biggest thing that's ever happened.

And the attempt to make it right usually makes things worse.

Sometimes the loops are almost comical. Versace is building a new hotel in Dubai, for instance, but the beach sand now gets so hot that guests burn their feet. Solution: a "refrigerated beach." As the hotel's founder explained, "We will suck the heat out of the sand to keep it cool enough to lie on. This is the kind of luxury top people want."[127]

Sometimes it's not shake-your-head funny but almost unavoidable. As more and more of Australia desertifies, the country could find itself "using 400 percent more energy to supply its drinking water by 2030 if the policy trend towards seawater desalination were to continue."[128]

And often—usually in the poor world—it's simply tragic. "Drinking water in Bangladesh is often full of salt as rising sea levels force water further inland," a Dhaka newspaper reporter wrote recently. That means women have to trek ever farther for a pitcher of clean water—sometimes several trips of several miles a day. "Some reports claim women and adolescent girls no longer have enough time and energy to carry out household duties like cooking, bathing, washing clothes and taking care of the elderly and infirm. It is even affecting their marriage prospects and family lives. Families who struggle to get clean water don't want daughters to leave their homes and marry elsewhere." Adolescent girls forced to drink increasingly saline water found their skin was "turning rough and unattractive," and "men from outside the area had no interest in marrying them."[129]

That's life on our new planet. That's where we live now.

· 2 ·

HIGH TIDE

New planets require new habits. If you walk out the airlock on your Martian base and start breathing, you'll be sorry. If you find yourself on Pluto, a strong leap will take you 116 feet into the air. We simply can't live on the new earth as if it were the old earth; we've foreclosed that option.

In the world we grew up in, our most ingrained economic and political habit was growth; it's the reflex we're going to have to temper, and it's going to be tough. Across partisan lines, for the two hundred years since Adam Smith, we've assumed that more is better, and that the answer to any problem is another burst of expansion. That's because it's worked, at least for a long while: the lives of comfort and relative security that we Westerners lead are the product of ten generations of steady growth in our economies. As Larry Summers, now President Obama's chief economic adviser, put it while he was still Bill Clinton's treasury secretary: we "cannot and will not accept any 'speed limit' on American economic growth. It is *the task* of economic policy to grow the economy as rapidly, sustainably, and inclusively as possible."[1]

In a previous book, *Deep Economy,* I argued that for the rich nations growth is a dubious prescription, no longer delivering the psychic satisfaction it once promised. And the math has always been daunting, anyway: if the Chinese, say, ever owned cars at the same rate as Americans, the number of vehicles on the planet would go from 800 million to almost 2 billion; if they ate as much meat as we do, they would require two-thirds of the planet's grain harvest.

But now—now that we're stuck between a played-out rock and a hot place—it's time to think with special clarity about the future. On our new planet growth may be the one big habit we finally must break.

I understand that this is the worst possible moment to make such a point. The temporary halt to growth that we call a recession has—in an economy geared only for expansion—wrecked many lives. We're deep in debt, as individuals and as nations, and in an effort to climb out from underneath that economic burden, we've bet yet more money that we can get growth rolling once more. That's what an "economic stimulus" is—a wager that we can restart the growth machine and make back not just the amount we spent stimulating but the debt that caused the trouble in the first place. When Barack Obama offered his first budget early in 2009, critics immediately assailed it for driving up the nation's debt. The United States, according to estimates from the Congressional Budget Office, would "run budget deficits approaching $1 trillion every year for a decade under Obama's plan.[2] "We cannot have debt pile on top of debt," said one member of the Senate Budget Committee.[3] The president responded that the country was already running huge deficits to pay for follies like the war in Iraq. His new spending, on energy, education, and health care, would be different; it would put us on "a pathway to growth" that over time would shrink those deficits. "Let's make sure that we're making investments that we need to meet those growth targets," the president said. "It's going to be an impossible task to balance our budget or even approximate it if we are not boosting our growth rates."[4] The White House forecast that the size of our economy would be growing 4.6 percent by 2012.[5]

Far worse, of course, is the ecological debt we face—the carbon accumulating in the atmosphere and reshaping the planet. And there, too, the most obvious way out is a new round of growth—a giant burst of economic activity designed to replace our fossil fuel system with something else that will let us go on living just as we do now (or better!), but without the carbon. Even, or especially, as our economy has tanked, we've seized on the idea of green growth as the path out of all our troubles. Former vice president Al Gore and UN secretary-general Ban Ki-moon cowrote an essay for the *Financial Times* arguing that "we need to make 'growing green' our mantra"; they called for an international agreement to provide massive investments, "a green light for green growth."[6] The venture capital firm where Gore works part-time, Kleiner Perkins, has turned its attention from information technology to "clean-tech," offering investors a half-billion-dollar "Green Growth Fund." In Asia, regional leaders endorsed a "Green Growth" plan; "we have no doubt that the Chinese government has adopted a green growth model, though it may bear another name in China," one UN official enthused. (Korea actually came up with the best moniker for its green growth scheme: "Save our Seoul.")[7]

The most fervent evangelist for the green growth spurt (albeit a late convert) is Thomas Friedman, the *New York Times* op-ed columnist who serves as a kind of political GPS unit, always positioned just far enough ahead of the curve to give readers the sense that they're in the know, but never beyond the comforting bounds of conventional wisdom. After two best sellers on globalization (*The Lexus and the Olive Tree* and *The World Is Flat*) that somehow managed never even to discuss global warming, his most recent tome, *Hot, Flat, and Crowded,* suddenly seized on it as the greatest of our crises. But because the modern world is a "growth machine" that "no one can turn off," he rejects any real recalibration. Instead, he outlines a "Code Green" platform, which involves "inventing a source of abundant, clean, reliable, cheap electrons, which would enable the whole planet to grow in a way that doesn't destroy its remaining natural habitats."[8] He talks about plug-in hybrid cars in every garage and a "Smart Black Box" in every basement to track the energy use of every appliance in the home. If we do these things, he assures us, all will be well: "In such an America, birds will surely fly again—in every sense of that term. Our air will be cleaner, our environment will be healthier, our young people will have their idealism mirrored in their own government. . . . America will have its identity back, not to mention its self-confidence, because it will again be leading the world on the most important strategic mission and values issue of the day." He sounds like a man delivering an acceptance speech at the Democratic convention. When he finishes with a flourish—"So I say we build windmills. I say we lead!"[9]—you can almost see the balloons dropping. I'd vote for him.

And I'm glad, too, that President Obama is singing from the same hymnal. Given the political environment in which we operate, it represents a heartening advance over "Drill, Baby, Drill." Friedman is right: "We don't just need a bailout. We need a reboot. We need a build out. We need a buildup. We need a national makeover."[10] In fact, Friedman is helping to shape what has quickly become the conventional wisdom, the kind of thing that important people tell each other at Davos and on C-Span. Here's Larry Brilliant, the president of Google, announcing the company's very useful initiatives for clean energy in 2008: "We have chosen them both because we think solving them will make a better, fairer, safer world for our children and grandchildren—and the children and grandchildren of people all over the world."[11]

As usual, though, *grandchildren* is the tip-off. Smart people are starting to understand the size of the problem, but they haven't yet figured out the timing; they haven't yet figured out that the latest science shows that this wave is already breaking over our heads. So Friedman insists that what we need to do is build a technologically advanced America, a shining green city on a hill, and then the Chinese will emulate us. He wants the full array: the big windmills strung across the Midwest, the concentrated solar arrays in the Arizona desert, the thousands of miles of high transmission lines to connect them all. The longest soliloquy in his book is a hymn to the soon-to-be-smart home, where the solar panels call up to inform the utility when there's been a blackout, where the smart lights in your office are triggered by motion sensors, where you "plug your 'smart card' (sponsored by Visa and United Airlines Mileage Plus)" into your Sun Ray computer terminal to start your workday, where your house is so savvy that it knows "when the sun is shining brightly and the wind is howling" and automatically turns on your dryer to finish your laundry.[12] It will all be so appealing that we won't need to bother with any treaties or anything. "A truly green America would be more valuable than fifty Kyoto protocols," he says. "Emulation is always more effective than compulsion."[13]

If we had started twenty years ago, when we first knew about global warming, and when we had the first hints of peak oil, such a plan might have made sense. (Well, maybe not *utter* sense. After all, if the wind is blowing and the sun is shining, you could always dispense with the smart house and put your clothes on a fourteen-dollar clothesline.) The Chinese might have developed very differently if they'd had a better American example to look at. But we didn't do it twenty years ago, precisely because it would have interfered with economic growth. And now—well, now we're in the middle of the trouble. The waves are already breaking over the levee; the methane is already seeping out of the permafrost; the oil wells are already coming up dry. It's going to be a little late.

So, for the record, I support a green Manhattan Project, an ecological New Deal, a clean-tech Apollo mission. If I had money, I'd give it to Al Gore to invest in start-ups. These are the obvious and legitimate responses of serious people to the most dangerous crisis we've ever encountered, and to a real degree they're working. In late May 2009 the Energy Information Administration predicted that by 2012 wind power would produce 5 percent of the nation's electricity, which represents astonishing growth.[14] We really do need to cut carbon emissions by 40 percent by 2020, or produce all our electricity from renewable sources within a decade, or meet all the other targets that good people have identified. They are precisely the way our system should respond. And in large measure that's how it *will* respond. The next decade will see huge increases in renewable power; we'll adopt electric cars far faster than most analysts imagine. Windmills will sprout across the prairies. It will be exciting.

But it's not going to happen fast enough to ward off enormous change. I don't think the growth paradigm can rise to the occasion; *I think the system has met its match.* We no longer possess the margin we'd require for another huge leap forward, certainly not fast enough to preserve the planet we used to live on.

That is a dark thing to say, and un-American, so I will try to make the case carefully.

In the first place: this kind of transformation is a big job. Even in normal times, even on the old planet, the transition from one source of energy to another took many decades. Vaclav Smil, a Canadian scholar who has provided some of the most detailed analyses of our energy predicament, recently tried to calculate how long such shifts have taken in the past. Industry began burning coal in the mid-1700s, he points out, but it wasn't until 1892 that the United States burned more coal than wood (and it was not until the mid-twentieth century that Asia passed this threshold). Commercial production of oil began in the 1860s, but it took fifty years before oil captured a tenth of the global energy market, and then thirty more years to go from a tenth to a quarter. "Analogical spans for natural gas are almost identical: approximately fifty and forty years," Smil observes. So we shouldn't be surprised at how off more recent predictions have been. In the early 1980s, some advocates of alternative energy predicted that America would draw 30 to 50 percent of its energy from solar power, wind power, and biofuels by the first decade of the twenty-first century. Instead, those sources produce only about 1.7 percent of our energy today.[15] We don't have fuel-cell cars. We do have hybrids—we've had them for a decade—but in 2009 they comprise about 3 percent of the vehicles sold in the United States, even with generous federal tax incentives.[16] It's worth remembering as well that more than 75 percent of the power plants expected to be in use in 2020 have already been built.[17] Also, the big oil companies made two-thirds of a trillion dollars in profit during the Bush years, but in 2008 they invested precisely 4 percent of their winnings in renewable and alternative energy.[18]

Now, you could argue that we'll build a strong enough political movement to make those numbers move much faster; that's what many of us have spent much of the last decade trying to do, and with at least a little success. (That should help. There was, after all, no *crusade* to move us from wood to coal or from coal to oil.) And lately, at least in the United States, we've found some new supplies of natural gas, which is a good "bridge fuel" between dirty coal and clean sun—you can retrofit your coal-fired plant to burn the stuff. On the other hand, we use so much more energy now that the task is harder than ever before. When we took all those decades to go from wood to coal, the planet used the energy equivalent of about half a billion tons of oil each year. Today that number is about 9 billion tons. So even converting half of it to renewables is, as Smil points out, "a task equal to creating *de novo* an energy industry with an output surpassing that of the entire world oil industry—an industry that has taken more than a century to build." It's true that when we shut down the auto industry at the start of World War II, we built lots of airplanes quickly, but it's also true that the Liberty ships, "the ships that won the war" by carrying matériel and troops to Europe and Asia, dated back to the turn of the century.[19] It's true that in a decade we were able to land three men on the moon—but this time we're talking about sending all of us into orbit. "The historical verdict is unassailable," writes Smil. "Because of the requisite technical and infrastructural imperatives and because of numerous (and often entirely unforeseen) socio-economic adjustments, energy transitions in large economies on a global scale are inherently protracted affairs."[20]

For the dry academic term *socio-economic adjustments*,

substitute the phrase *sunk costs*; it's a phrase we need to know if we want to understand why all the big companies are not jumping aboard the clean energy train. The journalist Paul Roberts figured earlier this decade that "the existing fossil fuel infrastructure, from power plants and supertankers to oil furnaces and SUVs," is worth at least $10 trillion, and scheduled to operate anywhere from ten to fifty more years before its capital costs can be paid off.[21] If we shut it down early, merely to save the planet, someone will have to eat that cost. Given such "serious asset inertia," no owner or investor in a power plant is likely to accept the writedown without a "nasty political fight." Indeed, Roberts points out, we've already had such a donnybrook. The Clean Air Act, passed decades ago, would have required coal-fired power plants to stick expensive scrubbers on their stacks to keep mercury and sulfur out of the air. Instead, the utilities were politically strong enough to win exemptions for existing plants. They asked for, and they got, a stay of execution that remained in effect until the Obama administration.[22]

In case you think the fossil fuel companies are less powerful now than before, think again: Exxon Mobil made more money in 2006, 2007, and 2008 than any company in the history of money. "It's the world's greatest company, period," one Goldman Sachs analyst gushed to a reporter. "I would put Exxon up against any other company at any other period in time." Exxon has spent the last decade underwriting an elaborate disinformation campaign to sow doubt about climate change and with reasonable success; 44 percent of Americans believe global warming comes from "long-term planetary trends" and not the pumps at the Exxon station.[23] And the company has a clear idea of where its future lies. By its calculation, solar, wind, and biofuel will account for just 2 percent of the world's energy supply by 2030, while oil, gas, and coal will represent 80 percent of the pie—and Exxon and its ilk may possess the political power to make that a self-fulfilling prophecy. The company doesn't see "much business sense" in investing in solar or wind or geothermal. "For the foreseeable future—and in my horizon that is to the middle of the century—the world will continue to rely dominantly on hydrocarbons to fuel its economy," insists CEO Rex Tillerson.[24]

There are ways to make the transition happen faster than in the past. Almost every environmentalist around the world is working to raise the cost of fossil fuel, in the hope that higher prices will accelerate the switch. A cap on carbon, if it can be passed over the objections of the oil and coal companies, would make gasoline and electricity more expensive, sending the signal we need to change our habits—and thus spur the new wave of investments in clean energy. But again, that works only if the price of energy rises *enough* to take some existing piece of hardware out of service. Everyone has to keep voting for politicians who will raise the price of gasoline high enough to cause most of us to park our cars and take the bus. For that to happen, we'd need to build a movement more powerful than the energy industry, powerful enough to raise the price of coal to the point where energy companies will simply swallow their investments and start shutting the plants down. This goal won't be easy to accomplish. In early 2009, just as Obama was getting set to unveil his energy plans, word came that 2,340 lobbyists had registered to work on climate change on Capitol Hill (that's about six per congressman), 85 percent of them devoted to slowing down progress. The American Coalition for Clean Coal Electricity spent more than any other organization in Washington lobbying on climate change—and also producing a series of commercials, including one in which lumps of coal sing Christmas carols. Its goal: "robust utilization of coal."[25] And its work was made easier by our economic collapse. The amount of new installed wind capacity in the United States, for example, was actually projected to decline in 2009 from 2008, and T. Boone Pickens canceled plans to build the world's largest wind farm on the Texas Panhandle due to financing difficulties. "You've got an industry hanging on by its fingernails," said the chief executive of the American Wind Energy Association.[26] (Though by year's end orders seemed to be reviving.)

I routinely give speeches about global warming, and so I know from experience that one of the first three or four questions will be about nuclear power. A man—always a man—approaches the microphone and asks with barely concealed glee if building more reactors isn't the "solution" to the problem. His thought, usually, is that I am an environmentalist, and hence I must oppose nuclear power, and hence aren't I a moron. Which I may be, but in this case nuclear power mainly serves to illustrate the point I'm trying to make about the difficulty of changing direction quickly. It's quite true that nuclear power plants don't seem as scary as they did a generation ago—not that they've gotten safer, but other things have gotten nastier. I mean, if a nuclear plant has an accident, it's bad news, but if you operate a coal-fired plant exactly according to the instructions, it melts the ice caps and burns the forests. Still, nuclear plants *are* frightening, in part because new ones spill so much red ink. A series of recent studies have found that the capital costs of new conventional atomic reactors have gotten so high that, even before you factor in fuel and operations, you're talking seventeen to twenty-two cents per kilowatt hour—which is two or three times what Americans currently pay for electricity.[27]

And that's if the plant gets built on time. "Delays would run the costs higher," as one study put it, and nuclear plants are always delayed.[28] Consider, for instance, what happened in Finland, where the country (thinking ahead, in a Scandinavian way) decided in 2002 to build a new nuclear power plant in an effort to cut its carbon emissions. The *New York Times* called the choice *prescient,* and the right-wing Heritage Foundation heralded it as *rational,* but a more accurate adjective would have been *pricey.* It was supposed to be completed in 2009 but now won't be online until at least 2012, and the original budget has gone up by more than half to $6.2 billion. A reporter visiting the site found that the interior of the containment vessel "was lined with a solid layer of steel that was crisscrossed with ropy welds. On this surface someone had scrawled the word 'Titanic.'"[29] As a result of troubles like that, a 2008 report from Moody's Investors Services concluded that any utility that decided to build a reactor could harm its credit ratings for many years. A Florida utility, in fact, predicted that even a six-month delay in its building plans could add $500 million in interest costs. And this was all before the great credit crunch at the end of the Bush administration.[30] Bottom line: building enough conventional nuclear reactors to eliminate a tenth of the threat of global warming would cost about $8 trillion, not to mention running electricity prices through the roof.[31] You'd need to open a new reactor every two weeks for the next forty years and, as the analyst Joe Romm points out, you'd have to open ten new Yucca Mountains to store the waste.[32] Meanwhile, uranium prices have gone up by a factor of six this decade, because we're—you guessed it—running out of the easy-to-find stuff and miners are having to dig deeper.[33]

We can do this same kind of analysis all day, for every other energy source. Sometimes the news is just as bleak: the cost of retrofitting coal-fired power plants to catch the carbon dioxide coming out the smokestack is so high that almost no private enterprise is even trying to build such a plant, never mind the millions of miles of pipe needed to push the carbon dioxide back underground, an infrastructure project that Vaclav Smil has calculated would need to be as large as the oil infrastructure we've built up over a century. Sometimes the news is a little better: proponents insist, for instance, that new "fourth generation" nuclear reactors can be built quickly and cheaply. (And eat nuclear power waste to boot!) More strikingly, the cost of wind and solar power continues to decline fairly steadily, as we build more turbines and solar panels. And the last year has seen new discoveries of natural gas in the United States that could help wean us off dirtier coal. Still, "solving" about a ninth of the global warming problem would require 2 million large windmills; we'd need to build four times as many as we built in 2007, every year for the next forty. Doable, perhaps, but again that would only get us a ninth of the way—to 450 parts per million, which is already way too high.[34] Wherever we turn, we always bump our heads against the same bottom line: it's expensive, and it takes a long time to even try to replace our fossil fuel system.

And that's on the old planet. What we need to talk about now is what it's like to make massive change on the new one, where we're suddenly running out of fossil fuel and dealing with a spooky, erratic climate.

Begin with the most boring word in the political lexicon: *infrastructure.* It doesn't arouse passion—it's not like gay marriage, where everyone enjoys the pleasure of knowing they're right. And yet infrastructure—our physical stuff, our housing stock and our roads and our rail lines and our ports and our fiber-optic cables and our pipelines—is what defines us as an advanced economy. We've let ours deteriorate, of course. Even on the old, stable planet we were falling hopelessly behind. Today, one in every four bridges in the United States needs major repairs or upgrades. (The press conference to announce this finding was held under an interstate span near North Philadelphia where a six-foot crack in a concrete support pillar beneath I-95 shut down the highway for three days in 2008, choking secondary roads with 185,000 detoured cars and trucks a day.)[35] Compare the cost with, say, the bank bailouts: the same week we invested $350 billion to keep Citibank and its kin from failing, an official of the National Surface Transportation Policy Study Commission said that we'd need to spend as much as $225 billion in additional money every year *for decades* to avoid "the kind of gridlock that's coming that could drag our economy down with it."[36] If we don't act, we'll have "the infrastructure of a Third World country within a few decades," added an official of the Regional Plan Association.[37] Most Americans haven't traveled in Third World countries, so I'll clarify: that means potholes so big that you travel fifteen miles an hour if you travel at all.

Now add on global warming. Take the example with which I began this book, the smallest possible example you can find: the road that connects my small town to the rest of civilization. Route 125 is a state scenic road; it winds down the bank of the plunging Middlebury River, with gorgeous views at every season. But five times in the last twenty years that river has flooded, the two worst being those record storms in the summer of 2008. You can go to YouTube and see a video—the water washing over the road, tearing out pavement, nearly swallowing up a car that quickly retreats. It took the state of Vermont weeks of work to make the road safe again—dumptrucks by the score ferrying loads of riprap to shore up the banks, new guardrails and blacktop. It cost $1,164,566.77 to repair the road after those floods, which is real money when you remember that 550 of us live at the top of the hill. (It cost far more to repair our town roads. We'll be paying off the bond for many years, probably well past the next set of storms.) The bridge that washed out at the bottom of the gully was gone for months, and the people who lived on the wrong side added half an hour to their commute to school or work.

Here's what happens: a river that has to carry more water takes up more space. There's a mathematical formula for it, which shows that Route 125 doesn't really work anymore. The river it's running next to is getting wider because we now live on a planet where warmer air holds more water vapor and hence we have bigger storms. There's no room for the road to shift—it's carved out of the mountainside already. To keep our town connected, we're probably going to need a new road, up four or five miles of steep hill, at a cost, according to state engineers, that would begin at $2 million a mile. Which isn't going to be easy to find, because 43 percent of the state's roads are already in "poor condition" even without this new kind of damage, and the state's budget for road repair is already far below what everyone agrees is needed.[38]

As I say, that's the smallest possible example—important to me and the others who live in our town on top of the mountain, but literally one story in a million. Consider that same rainstorm in 2008, a storm that didn't make the national news. Just a big rainstorm. But a couple of hundred miles north of us, in Montreal, it overloaded an aging sewage system, flooding hundreds of homes and business. The storm came not long after a series of internal reports by Canadian government agencies warning that "climate change was threatening critical infrastructure across the country," and that urban water infrastructure in particular "has the potential to suffer the greatest damages or losses associated with climate change unless proactive adaptation actions are taken." They weren't, obviously, because—well, because retrofitting a nation's sewer system to deal with this new planet is expensive. (The U.S. federal government says it will take $500 billion-with-a-*b* to fix America's sewers over the next twenty years;

in 2008 it allocated $687 million-with-an-*m*.)[39] But it's expensive when raw sewage flows into your basement, too. After that August storm one Montreal businessman announced that he was suing the city to recover his $400,000 in losses from just one warehouse that flooded.[40]

So now try to multiply that $1.1 million for our new road, or that $400,000 for one guy's basement full of crap, by all the change that happens when the infrastructure of a new world, already aging and breaking, meets the new world we've built. Sometimes it's huge and traumatic and obvious; after Hurricane Katrina the federal government ponied up $130 billion for repairs, which, as anyone who has visited Louisiana can attest, was Not Enough. And fighting off the next Katrina will be far more expensive. By 2011 the federal government will have spent about $14 billion rebuilding levees to protect New Orleans from "moderate-sized" hurricanes. But if you wanted to keep that one city safe from a Category 5 hurricane, you'd need $80 billion and another twenty years of work. And that would do the trick only if you were talking about the old Category 5 hurricanes, in a world with sea levels about where they are now. Sea level rise "will add substantial cost and complexity" to the job, officials reported with characteristic understatement. Indeed, they'll need to raise floodwalls at least three feet. And since the Army Corps of Engineers, by law, can only plan projects to last fifty years, that's all the sea level rise they'll build into their work, even though every climate model that we have now says the oceans will be rising for centuries to come.

"There is not enough money, and there is not enough mud, to do everything we want to do," a Louisiana official reported. And yet if you don't spend the money, you still spend the money: even once the current levee projects are done, officials estimate that a future Katrina-strength storm will do damage of $152.8 billion.[41] The rising sea is already turning the town of Port Fourchon into an island. A new $538 million elevated highway is under construction to keep it connected, which seems like a lot of money to spend for a small town, except that 18 percent of America's oil comes ashore through this small town—five thousand of the planet's six thousand offshore drilling rigs are located nearby in the Gulf of Mexico.[42] And all that oil? Burning it releases that much more carbon, meaning that much higher ocean, meaning—oh, whatever.

If the cost of repairing hurricane damage is high in places like New Orleans, consider places where there's no real hope of finding the money. A year after Cyclone Sidr ravaged Bangladesh in 2007, a million Bengalis were still without homes. People were given plastic sheeting after the storm, but a year later "the sheets were torn," according to a government official.[43] In the Pacific islands, cleanup costs for Cyclone Heta cost the nation of Niue five times its annual government budget and more than its total gross domestic product, chasing away so many residents that it may soon drop below the population threshold required by the United Nations for membership.[44] The string of hurricanes that wracked Haiti in 2008 destroyed 60 percent of the country's harvest "and entire cities were rendered desolate and uninhabitable," but international donors, "distracted by the credit crunch," came up with barely $30 million to help, the *Guardian* reported. "At the current rate of clean-up," the newspaper went on, "it will be almost three years before the mud and debris is cleared."[45] In Cuba, the same train of storms caused at least $5 billion in damage and destroyed or damaged a staggering 450,000 homes, and a third of the nation's crops were lost.[46]

The new planet we live on is inherently more expensive than the old one. The wind blows harder; more rain falls; the sea rises. It would cost more to settle it if we were just arriving from outer space, but the real price tag comes because we built it up so thoroughly during our ten thousand stable years and now must defend that investment. These are sunk costs, and they're in danger of being *sunk* costs. Consider Venice, plagued by flooding as it sinks and the ocean swells. The city, in 2011, will finally finish a long-planned series of barriers, the "Moses project." At a cost of $5.5 billion, it will protect the city against sea level rise expected through the end of the century.[47] Planning has already begun for what would be America's longest seawall, the $4 billion "Ike Dike" to protect Galveston, Texas.[48] The Netherlands will need to spend at least 100 billion euros on dike upgrades, and another 300 million dumping enough sand to expand the North Sea coast by a kilometer.[49] Hilton Head, in South Carolina, has already spent $60 million this decade on "beach nourishment." Jill Foster, deputy director of community development for the town, said, "People here have blinders on and think sea level rise is two centuries out. . . . Frankly, we can't guarantee we can always renourish the beach. I mean, look at this economy."[50]

In its waning days even the Bush administration began to admit a little fear. In the spring of 2008 the U.S. Department of Transportation issued a study on the effects of a twelve-inch rise in sea level along the Atlantic Coast, a figure most scientists would now dismiss as laughably small. Even so, it predicted that such a rise would be enough to "frequently" flood the Capitol and the Lincoln Memorial. Rail lines and airports would have to be moved—and "about a quarter of homes and other structures within 500 feet of the U.S. coastline . . . will be overtaken by erosion during the next sixty years."[51] Israel faces flooding along the entire coastline, "causing irreparable damage to essential infrastructure like ports and power plants"; one study put the cost at $33 billion a year.[52] A two-foot rise in sea level "would make life in South Florida very difficult for everyone," a recent study concluded.[53] "A large belt around the tip of Manhattan—including Wall Street—would have a 10 percent chance of flooding in any given year" with even modest rises in sea level, a recent report found.[54] Does that sound implausible? Even on the old planet, New York was vulnerable: a stiff nor'easter in 1992 sent a storm surge eight feet above sea level, flooding the FDR Drive with four feet of water and shutting down LaGuardia Airport. The estimated price tag for one day's disruption: about a quarter billion dollars.

The list goes on endlessly, one place after another, one billion after another. In Alaska, infrastructure costs will rise about 20 percent between now and 2030 because of global warming, one study concluded, but its author added that the figure, based on a survey of state agencies, was likely an underestimate. "On more than one occasion I had people laugh at me on the phone," he said.[55] But probably not in the village of Newtok, on the Ninglick River on the state's western coast, which offers some small sense of how much it costs to simply give up trying to defend the old planet and move people to safety. The U.S. military says the evacuation of Newtok will be good practice; a marine contingent from New Orleans will supervise the transfer of 250 people and the construction of sixty homes. At a total cost of $130 million, or about $400,000 per person.[56] Do the math. There's not enough money on earth, not even close, to evacuate every town threatened by rising sea levels. At some point, it's all good money tossed after bad: in the spring of 2009, even as the Obama administration started doling out stimulus money, a team of scientists reported that significant sums were going for highways, housing, and schools that would soon be underwater—like Highway 87 on Texas's Bolivar Peninsula, which models show will eventually be submerged in many places, "leaving the dry sections accessible only by boat."[57] If you want to know who *is* planning ahead, here's a story from the Australian papers in June 2009:

new construction plans for the world's largest coal export facility had been quietly altered to raise the structure two or three meters for fear of rising sea levels.[58]

But the direct costs of moving people or building dikes may be the least of the expense. Let's think for a moment about a technology that gets little attention but provides an essential foundation for our prosperity. Not the power plant; the actuarial table. It's a remarkable invention: by looking at past deaths, or fires, or floods, or crop failures, or knee injuries to fullbacks, actuaries can reckon the chance of such events in the future. That enables them to underwrite insurance at a reasonable cost—and that insurance lets us do everything else. Who would build a house without it, or a factory? (That's why insurance is by some measures the world's largest industry.)[59]

The art of underwriting is now highly complex and computerized. The day before Hurricane Gustav hit the Gulf Coast, for instance, models were forecasting that it would cause exactly $29.3 billion in property damage, and that its fury could destroy 59,953 buildings.[60] But that kind of precision masks the one huge flaw of the actuarial table: the technology is dependent on the planet behaving in the future as it has in the past. If we switch planets we need new actuarial tables, and we don't know what to base them on. Insurance payouts have been skyrocketing for the last decade, and they'll keep going up. In areas with frequent storms, the Association of British Insurers recently predicted 100 percent premium increases for policyholders over the next ten years.[61] But that's the good case: premiums would rise, just like seawalls, and it would cost money and be a drag on the economy; still we could make incremental adjustments. What we can't afford is the cost of complete uncertainty—or, rather, the cost of certainty that we're going somewhere new and unstable. What if you were selling life insurance and suddenly there was a global outbreak of some new and deadly plague? You'd be out of luck, not to mention out of business. "What we have seen in recent years in terms of insurance losses are but a harbinger of things to come," said Tim Wagner, cochairman of the Climate Change and Global Warming Task Force for the National Association of Insurance Commissioners. "Insurance is priced based on statistics and probability. What climate change has done is create ambiguity and uncertainty in the pricing scenario."[62]

Swiss Re, the world's biggest insurance company, wanted to figure out some of these possibilities, so it contracted with Harvard's Center for Health and the Global Environment for a report on the most likely outcomes, which was published in 2005. The Harvard team modeled two "climate change futures," one with the kind of gradual change we used to expect, and the other with the kind of disruptive, quick, and nasty change we've already seen. (The team didn't even bother modeling a worst-case scenario—"slippage of ice sheets from Antarctica to Greenland, accelerated thawing of permafrost with release of large quantities of methane"—that comes closest to what we're experiencing on the new earth.) Even in the milder scenario, climate change "threatens world economies." But their second, more real-world simulation predicts that as storms and other disruptions become more frequent, they "overwhelm the adaptive capacities of even developed nations; large areas and sectors become uninsurable; major investments collapse; and markets crash." Pay careful attention, despite the bland phraseology: "In effect, *parts of developed countries would experience developing nation conditions for prolonged periods* as a result of natural catastrophes and increasing vulnerability due to the abbreviated return times of extreme events."[63]

Since these are the words of people who write insurance policies for a living, let me translate: if you get sucker-punched by one storm after another, you don't have time to recover; you spend your insurance payout reroofing your house, and then the roof blows off again the next year. Maybe your insurance company cancels your policy (as has already happened this decade to millions in storm-prone coastal areas), and after the next storm or two your town starts looking less like America and more like Haiti. Meanwhile, the business that employs you loses its warehouse two years in a row, and then the insurance company either cancels its policy or jacks the rate up so high that it shuts down. Maybe the government becomes the insurer of last resort, as has already happened with flood insurance, but then the losses fall on all of us taxpayers, and we have to do with less funding for education or health care or, hmm, infrastructure. Between 2005 and 2007, state-run insurance programs in the United States saw their exposure *double* to $684 billion as people lost their private insurance.[64] The EU has set aside a billion euros a year for a "solidarity fund" to cover "uninsurable risk" to government-owned property, but new forecasts predict that floods alone will soon be doing 1.2 billion euros worth of damage to such facilities each year. "With a worsening climate, an increase in fund resources is needed," one bureaucrat said dryly.[65]

And if this is happening in the West, imagine the effect in poor countries: in their scenarios, the Harvard team reported, "the emerging markets are most hard hit, with widespread unavailability or pricing that renders insurance unaffordable. As a result, insurers withdraw from segments of many markets, stranding development projects."[66] This is not just speculation; a recent MIT study found that the GDP of poor countries dropped by 1 percent in those years when temperatures were a degree or more above average.[67]

Every feature of this new planet increases the uncertainty; we've already seen that coral reefs are dying off rapidly and could be gone altogether by midcentury. That's a tragic loss for the planet's biological diversity, and it damages the tourist industry on all the low-lying islands that are trying their best to cope with sea level rise. But it also removes the most important line of defense against storms on those coasts; one study suggested that a single kilometer of sheltering reef was worth $1.2 million.[68] Whole new categories of risk appear. As the number of thunderheads in the atmosphere steadily increases, so do the number of hailstorms. Australian insurers recently predicted that the number of storms with golf ball–size hail could become twice as frequent between now and 2050—which is no small thing since the third-most-costly natural disaster in Australian history was just such a storm that struck Sydney in 1999.[69] The total exposure of insurers is mind-boggling: in the five northernmost coastal counties of Texas alone, insurers are on the hook for $890 billion worth of risk, third in the nation behind Florida and New York.[70] And the costs are not confined to the coast. For me, standing by the bank of the Middlebury River, the single scariest statistic in the whole report may have been this: "A ten percent increase in flood peaks would produce one hundred times the damage of previous floods, as waters breach dams and levees."[71]

If it seems that I am callously reducing the danger we face to dollars and cents, that's correct. Money, in our system, equals information. It's how we understand risk; it's how we measure possibility; it's the only gauge we have for understanding our collective future. If you have a lot of money, you have a lot of options, and if you don't have much, your options narrow. On this new planet we'll have less money than we thought we would, and hence fewer choices.

You can see it happening already, in a thousand modest ways.

When the price of oil rose in 2008, for instance, the price of pavement—which is made from oil—went through the roof. Local officials relied on thinner pothole patches and "microseals" (which wear out faster), and they started repairing fewer roads.[72] "If you were doing ten roads in your town, you're only going to do five," said one upstate New York official.[73] A Tennessee mayor pointed out, "We can't afford $100,000 a mile. These are roads that school buses travel and emergency vehicles travel. All those roads are going to start to deteriorate."[74] How did we manage to push the price of oil back down? Only with a recession—with the longest siesta from growth since the Great Depression, and not just here but everywhere.

The sole even remotely plausible way out of this box canyon would be, as I've said, massive investment in green energy, but our mountains of accumulated debt make that harder, not easier. The current Australian government was elected on the promise of fighting global warming, but the economic slowdown quickly cooled its ardor, delaying the start of even modest plans for carbon reductions by two years.[75] In equally progressive New York, Democratic governor David Paterson quietly undercut the state's plan to control greenhouse gas emissions, a plan that had been crafted by a Republican predecessor, but in better economic times. "The governor was very concerned about undue cost on business," said one cheerful utility executive.[76] On the other side of the globe, the *Washington Post* reported, China was backing down on plans to close some polluting factories: "Money is increasingly needed to pay the salaries of workers whose companies have gone bankrupt and to provide social services for the rural poor, who are having trouble selling their crops. There has been less money left over for environmental initiatives."[77] China is building half the world's new floor space annually, but developers are mostly forgoing new green design techniques because the cost is now too high. In 2009 the Asia Development Council found that 80 percent of new construction is now of buildings that use "two to three times more energy than buildings in developed countries."[78]

As the recession began to build, China announced that some of its stimulus money would go to build more railroads. Which sounds, to our ears, very green: trains! Except that "among the upgrades money is destined for the construction of special lines for the transportation of coal between the major coal terminals of the country."[79] A month or so later, the Chinese government announced that in fact it planned to increase its coal production about 30 percent by 2015, which, in the words of former Clinton energy official Joe Romm, would be enough to "single-handedly finish off the climate."[80] When you're poor, you burn what's cheap. I'm on the volunteer fire company in our town, and one reason we dread high oil prices is that people start throwing anything they find into their woodstoves.

Dhaka, the capital of Bangladesh, sprawls across the lower delta of the Ganges River. It's a city without an accurate census but with a population somewhere in the neighborhood of 14 million residents (or nearly two New Yorks). We're used to thinking of Bangladesh as a "basket case," to use Henry Kissinger's pejorative, but in fact it's quite a miraculous place. It's incredibly crowded—150 million people in an area the size of Wisconsin—but they manage to feed themselves. It's a traditional Muslim society, but the average family has gone from having seven children to fewer than three in just thirty years. The bicycle rickshaw is the basic means of transit, and there are four hundred thousand rickshaw drivers in Dhaka alone. These Bangladeshis, in short, have done very little if anything to create this new Eaarth.

But they sure as hell live on our new globe. Before long the water that flows down the Brahmaputra and the Ganges will start to dry up, because the Himalayan glaciers that feed those rivers are melting. Meanwhile the Bay of Bengal is rising, pushing saltwater ever farther into the country's agricultural heartland. But when I was last there, in the summer of 2000, everyone was focused on a much smaller result of global warming: *Aedes aegypti,* a species of mosquito identifiable by white markings on its legs and a spot the shape of a lyre on its thorax. That summer, *Ae. aegypti* was triggering the first big outbreak of dengue fever that Dhaka had ever seen. The front pages of the newspapers carried giant sketches of the mosquito and stories like "Life-Saving Blood Needed for Young Poet Raquibul." They also carried the depressing news that with dengue fever, unlike malaria, bed nets were not going to help, unless you spent all day under the covers: *Ae. aegypti* dines from shortly after dawn till shortly after sunset. Not only that, but there's no vaccine and no real treatment save "close monitoring of vital signs." If you hemorrhage badly—sometimes from every orifice—your doctor will try transfusions.

I was spending a lot of time in the slums in Dhaka, and I got bit by the wrong mosquito. I was sicker than I've ever been, the kind of fever where sweat runs off your outstretched arm like rain off a gutter. But I was healthy going in and young and well fed, and so I didn't die. Plenty of people did: children, old people, weak people. As public health volunteers spread out around the city, they found that the mosquitoes were breeding in the jars that poor people used to store water. In old tires, too, but there weren't very many of those, since tires get repatched till there's nothing left of them.[81] The mosquitoes were also breeding in coconut shells, which prompted a public health campaign to turn them upside down. It didn't really work—dengue has come back to Bangladesh every year since. And not just Bangladesh or even Asia. As the *Economist* reported in 2007, "Fuelled by climate change, dengue fever is on the rise again throughout the developing world, particularly in Latin America. Mexico identified 27,000 cases of dengue fever last year, more than four times the number in 2001. In El Salvador, whose population is not much more than 6% of Mexico's, the number soared to 22,000 last year, a 20-fold increase on five years earlier. Uruguay recently reported its first case in 90 years."[82] A 2008 outbreak in Brazil was so bad that until the army opened a series of field hospitals, people were waiting more than a day even to be admitted to emergency rooms. "I am just watching my son die slowly as we knock on different hospital doors," one father said.[83] "Dengue has come to stay in Latin America," the Argentinian health minister said in 2009—after a year that saw more than a million cases across the region.[84]

The link to climate change couldn't be clearer: not only do warmer temperatures extend the geographic range of the mosquito (up to half the world's population is now at risk), but *Science Daily* reports that global warming "also reduces the size of *Ae. aegypti*'s larva and, ultimately, adult size. Since smaller adults must feed more frequently to develop their eggs, warmer temperatures would boost the incidence of double feeding and increase the chance of transmission. In addition, the time the virus must spend incubating inside the mosquito is shortened at higher temperatures. For example, the incubation period of the dengue type-2 virus lasts 12 days at 30 C, but only seven days at 32–35 C. Shortening the incubation period by five days can mean a potential three-fold higher transmission rate of disease."[85] But of course dengue fever is linked to poverty as well—you find fewer people storing water in open jugs in your average condo complex. Here's the real kicker, though, and the reason I even bring up dengue: economically, it functions like a hurricane. Having to treat it makes you poorer still, reduces your options even further. A 2009 Brandeis University study that looked at just eight nations, most of them small, found that dengue was costing them $1.8 billion annually.[86]

Consider Malaysia, a country whose economy has grown smartly in recent decades. But so have the number of Malaysian dengue cases. One recent study found that treating each victim costs $718, or one-fifth of Malaysia's per capita GNP, about 70 percent of which is paid by the government. That's the equivalent of fifty-three days of lost economic output for each person treated.[87] And in turn that's less money to spend on education, or HIV prevention, or infrastructure, or cleaning up the streets so that there's less dengue next year. In fact, Malaysia's health minister, Chua Soi Lek, placed the blame for the most recent outbreak squarely on the slum dwellers. "These selfish people like to blame others whenever there are dengue cases in their areas," he told the *New Straits Times*. "They blame the authorities for failing to collect rubbish on time and for stagnant drains."[88]

It's not just dengue, of course; malaria is also on the rise, and the *Washington Post* reported recently that "it's a near-certainty that global warming will drive significant increases" in cholera and other waterborne diseases. "Rainfalls will be heavier, triggering sewage overflows, contaminating drinking water, and endangering beachgoers. Higher lake and ocean temperatures will cause bacteria, parasites, and algal blooms to flourish."[89] In Lima, a heat wave in 1997 that "turned winter into summer" also led to an additional 6,225 cases of childhood diarrhea that cost the Peruvian health authorities an estimated $277,000 to treat.[90] (The number of cases goes up 8 percent for each one-degree increase in temperature, according to a 2009 Oxfam study.)[91] It doesn't just happen in poor places. U.S. authorities have blamed climate change for the spread of West Nile virus across the country, for instance, and a single storm dropping two inches of rain, the kind that's been increasing steadily, raises infection rates by a third.[92] But in the rich world we still have some margin; record rains in my town may wash out the road, but record rains in Mozambique not long ago washed out huge quantities of land mines that had been planted in nearby fields during the country's recent brutal wars. People swam into fields and died. To make matters worse, as one local official explained, "the government had spent thousands of dollars mapping the minefields just before the rains started last month. We had everything under control but they're all shifted now and some are even washing up on local beaches around the lake. Farmers are terrified to plough their fields again because they don't know what is under the silt."[93]

As that example implies, people's minds can be damaged as much as their bodies. Doctors in countries with bad heat waves report an upswing in psychosis. One young man was so bothered by the link between warming and drought that he convinced himself millions would die if he drank a glass of water. In the wake of Hurricane Katrina, the incidence of severe mental illness doubled in the affected areas, with 11 percent of the population suffering from post-traumatic stress disorder, depression, panic disorder, and a variety of phobias. After such a disaster, researchers report, "people feel inadequate, like outside forces are taking control of their lives."[94]

Which, actually, they are. It's an odd metaphor, but imagine the poor world as a man on a health club treadmill, working hard to keep pace with the machine. The rich world stands nearby, occasionally handing him some aid in the form of a PowerBar or a sweat towel—but also, now, pushing the button that speeds up the pace and steepens the grade. Global warming turns the idea of "development" into a cruel joke. We may wear wristbands expressing our commitment to "Millennium Development Goals" and the like, but in fact we're busily taking away what little security most people in the poor world actually have. The one asset a peasant always possessed—the confidence that if her grandmother grew corn or wheat or rice in a particular field, she could, too—erodes daily, as monsoons fail, floods roar, the sun bakes.

And the fact that so much of the world remains so poor is also one of the biggest obstacles to actually doing something about the climate. Just as we come into this crisis with an infrastructure deficit and an overhang of debt, so we also suffer from a justice deficit that will slow any attempt at action. In 2008, after a careful study of prices for goods and services in developing countries, the World Bank issued a new set of numbers: 1.4 billion people, it found, lived below the poverty line, 430 million more than previously estimated. What defines the poverty line? $1.25 a day. As the *New York Times* wrote at the time, "The poverty expressed in the World Bank's measure is so abject that it is hard for citizens of the industrial world to comprehend." (Indeed, a person living at the poverty line would be seventy-five cents short if he tried to buy a copy of the *New York Times*.) And the remedies? Exactly the things in ever shorter supply: India needs to increase farm productivity, all developing countries must "spend more on education," and Africa "requires stability, above all, to encourage investment."[95] Instead, when the market for biofuels exploded in 2008, we saw food riots in thirty-seven countries and as many as 100 million Africans moving back into poverty.[96]

So let's try to think how the new planet looks to someone living that kind of life. On the one hand, you probably realize that the weather is changing, and that it's causing real problems. On the other hand, you're very poor and eager for some small part of the comfort the rest of the world takes for granted. Only one in three Africans has electricity; in the countryside only one in ten can flip on a light switch.[97] In China, visitors immediately notice that everyone is forever spitting, hawking up great gobs of mucus. As Paul Roberts explains, "For generations, peasants have suffered near-constant respiratory infections—and thus have acquired the hawking habit—in no small part because they cannot afford to heat their homes in winter." The walls turn white with frost. "In my village," recalled one peasant, "when a girl was preparing to marry, the first thing the parents checked was, will the back wall of the would-be son-in-law be white or not? If not white, they approved the marriage because it meant his family was wealthy enough to keep the house warm."[98] If you want to have a light in your Congolese hut, or if you want to keep your Chinese child from having a chest cold that lingers all winter, the easiest thing to do is burn more coal. All the more so if your alternatives are disappearing. As one analyst pointed out in September 2008, "With its hydropower from the Himalayas drying up, India, even with its push for more wind and solar power, faces a rapidly-rising need for coal."[99]

In that world, how do you sit down and negotiate a global climate pact? No one has tried harder to game out the scenarios than Tom Athanasiou and Paul Baer, the directors of the Greenhouse Development Rights Network, who point out that this will be the first time that developing nations have ever come to international talks with any real clout; after all, if they burn all their coal, there's nothing the rest of us can do to ward off global warming. And they have justice on their side, since they've done nothing to cause the global warming. "The fact is, they're not particularly disposed to experiments that, they fear, will close off the only routes to progress they've ever had," they say.[100] In essence, poor countries would be required to sacrifice Plan A—the plan we've sold them on for decades, the Washington Consensus idea that they would get rich the same way we did. They've been obediently "reforming" their farming to look like ours, and shunting the displaced peasants off to the city to get jobs in factories producing cheap stuff, and trusting that these actions would "lift them out of poverty." In at least the case of China, the strategy has begun to work. But without cheap coal and cheap oil, the plan no longer makes sense.

So the obvious replacement for Plan A—for the now vain hope that the rest of the world can emulate us and messily grow its way into lives of relative comfort and security—is a Plan B, a grand bargain where the global North decides to *share* with the global South. And in return the South agrees to develop on a different, cleaner path. This has been the subtext of global climate talks since the very start, back in Rio de Janeiro in 1992. Reduced to the simplest terms, the deal goes like this: You give us enough windmills, and we won't burn our coal. You rebuild our factories so they're efficient, and we won't burn our coal. You come up with some other attractive ways to lift us out of poverty, and we won't burn our coal. Everybody knows that this will have to be the eventual bargain, and everyone has spent twenty years trying to game the talks.

As far back as the summer of 1997, when the original Kyoto treaty was being negotiated, the fossil fuel lobby asked its captive politicians to propose a "sense of the Senate" resolution that stated that the United States should not be a signatory to any protocol that did not include binding targets and timetables for developing nations as well as industrialized nations, or it "would result in serious harm to the economy of the United States." Everyone knew that such a restriction would doom the agreement, since the Chinese were producing a tenth as much carbon per capita as Americans and could not have been held to the same targets as were needed in the West. And even though Vice President Al Gore symbolically signed the document for the United States, it was never submitted to the Senate for ratification. President George W. Bush, of course, kept up the same line of argument. "I couldn't in good faith have signed Kyoto," he said in 2005, noting that the treaty didn't require other "big polluters" such as India and China to cut emissions. (He also added, ironically, that Kyoto would have "wrecked the economy.")[101]

The Chinese played Ginger Rogers to America's Fred Astaire, following our lead effortlessly with endless (accurate) invective about the "unshirkable responsibility" of the West for the climate mess. "It must be pointed out that climate change has been caused by the long-term historic emissions of developed countries and their high per-capita emissions," Foreign Ministry spokeswoman Jiang Yu said in February 2007, which was about the month that China finally produced as much carbon dioxide as the United States.[102] China's rapid conversion into a coal-burning machine may well be remembered as the biggest foreign policy failure of the Bush years, which is saying something. Instead of using every ounce of charm and pressure and incentive to deflect China's energy trajectory, we sat on the sidelines. In fact, we cheered the Chinese on.

In the time since the Bush administration finally decamped, both sides have tried to reengage. The Chinese realize they have a problem with climate change. By one account, weather-related disasters are already costing the country 5 percent of its GDP each year.[103] That's a lot—for every twenty yuan the economy produces, one goes to pay for freak rainstorms and severe droughts, which are growing rapidly. A Chinese government report in 2008 also predicted that climate change will cause China's production of wheat, corn, and rice to drop by as much as a third over the next fifty years.[104] That's a lot too—take every third basket of rice and toss it in the sea. So China's rhetoric has begun to change, and its interest in some kind of agreement has clearly grown—but only if the West ponies up. China's chief meteorologist admitted recently that heat waves and other disasters stemmed from climate change, but added that if China is to completely transform its energy structure and use clean energy, "it would need a lot of money."[105]

Some of his colleagues tried to put a dollar figure on "a lot" at the international climate talks in December 2008, saying that China wanted "rich countries to commit to donating one percent of their gross domestic product to help poor countries fight global warming."[106] That *is* a lot—it's hard to imagine explaining to the average American voter that he's going to work a couple of hours a month to donate windmills to China. In fact, Senator John Kerry, who at the time was about to take the reins of the Senate Foreign Relations Committee, immediately said that American economic problems would make that kind of assistance difficult. Our economic problems, he said, meant that the United States "is not going to be in the position we were two years ago, in the short term to do as much technology transfer or other economic assistance" to China. But he was quick to add that America's reluctance "shouldn't change the willingness of some other cash-flush countries" to chip in.[107] In fact, most rich nations were balking. "There is now a question mark hanging over the impact of financial crisis on making available new money" to poor countries, said one UN diplomat, diplomatically.[108] Meanwhile, the price tag keeps going up. In August 2009 a British study found that the costs of adaptation for poor countries would be three times higher than originally projected—as much as $500 billion annually once you added everything up.[109]

It's not that we won't reach some kind of agreement; we will, eventually. In fact, since Barack Obama's arrival, there's been real movement: we've passed the beginnings of legislation in the Congress and reached a preliminary deal at Copenhagen. But the kind of deal big enough to put us back on track toward 350 parts per million, the kind that had any hope of really altering our makeover of the earth? How likely is that? Meeting even minimal demands from the developing world would require increasing funding for adaptation eleven thousand times. The ranking Republican on the relevant House committee has already complained to the *Wall Street Journal* that "for a U.S. family of four, China's demand comes to nearly $1,900 in yearly taxes."[110]

If you want to understand the limits on our response, just listen to Obama. Here he is in February 2009 discussing calls for greater spending in his stimulus plan: "Let's not make the perfect the enemy of the essential."[111] And in July, on a call with Internet journalists about health care reform, he said that he refused to let "the perfect be the enemy of the good."[112] That same month, speaking about the upcoming Copenhagen climate talks, he said the same thing: "We don't want to make the best the enemy of the good."[113] It's sound and sane politics—in the first two cases. Because economic policy and health care are perfect examples of normal politics. You split the difference between positions, make incremental change, and come back in a few years to do some more. It doesn't get impossibly harder in the meantime—people will suffer for lack of health care, but their suffering won't make future change impossible. Global warming, though, is a negotiation between human beings on the one hand and physics and chemistry on the other. Which is a tough negotiation, because physics and chemistry don't compromise. They've already laid out their nonnegotiable bottom line: above 350 ppm the planet doesn't work. In this case, the good and the essential and the perfect and the adequate are all about the same.

So one can hope for unexpected breakthroughs. But one can also fear novel kinds of trouble. On the new world we've built, conflict seems at least as likely as cooperation. In 2006, British home secretary John Reid publicly fingered global warming as a driving force behind the genocide in Darfur, arguing that environmental changes "make the emergence of violent conflict more rather than less likely. The blunt truth is that the lack of water and agricultural land is a significant contributory factor to the tragic conflict we see unfolding in Darfur. We should see this as a warning sign."[114] When *Time* magazine's Alex Perry traveled to the region the following year, he reported that "the roots of the conflict may have more to do with ecology than ethnicity." The few pockets of good land had always been prized; as rainfall has decreased for the last five decades and the Sahara advanced, smothering grazing land with sand, "the competition is intensifying." In Darfur, "there are too many people in a hot, poor, shrinking land, and it's not hard to start a fight in a place like that."[115] Eight years of drought have also accelerated fighting in Somalia, while crop failures have made the misery in Zimbabwe ever worse. Many of the refugees who fled the expanding Sahara settled on the borders of massive Lake Chad, only to see its waters shrink by 90 percent since 1973.[116] In Syria, 160 villages were abandoned after a 2008 drought, and elsewhere in the Middle East the International Institute for Sustainable Development predicts that even modest global warming will lower the volume of the Euphrates River by 30 percent and will shrink the Dead Sea by 80 percent.[117] (A follow-up study in the summer of 2008 predicted that the "ancient 'Fertile Crescent' will disappear this century.")[118] A one-meter rise in sea level would obliterate at least a fifth of the Nile delta. Meanwhile, increased evaporation and new upstream demand seem set to reduce the river's flow by 70 percent. It's a region, said one observer in 2009, whose "death warrant may already have been signed."[119] In Kashmir, Indian and Pakistani troops have long faced off over the Siachen Glacier, at nineteen thousand feet the highest war zone on earth. But now the glacier is melting fast, leaving "not much left to fight over," in the words of the journalist Priyanka Bhardwaj, except of course the millions of Pakistanis who will be affected by a "severe water crisis" when it disappears.[120] Here's former British foreign secretary Margaret Beckett, speaking during the first-ever debate on climate change and armed conflict at the UN Security Council. "What makes wars start?" she asked. "Fights over water. Changing patterns of rainfall. Fights over food production, land use. There are few greater potential threats to our economies too . . . but also to peace and security itself."[121]

She's not alone in that assessment. Four major studies in the last two years from centrist organizations in the United States and Europe have concluded that "a warmer planet could find itself more often at war." Each report "predicted starkly similar problems: gunfire over land and natural resources as once-bountiful soil turns to desert and coastlines slip below the sea." The experts also expected violent storms to topple weak governments—which makes a certain amount of sense to those of us who watched George W. Bush begin his descent in the polls after he bumbled the response to Katrina.[122] "Billions of people would have to move" if temperatures rise four or five degrees, the British economist Nicholas Stern predicted recently.[123] The directors of climate research for the Center for Strategic and International Studies in Washington predicted recently that as "climate-induced migration" increased the number of "weak and failing states," terrorism would likely grow.[124] By midcentury, according to some recent models, as many as 700 million of the world's 9 billion people will be climate change refugees.[125]

What would such mass migration look like? Consider

Bangladesh: melting glaciers upstream, rising seas downstream, dengue spreading in the cities. It's the most densely populated large nation on earth, and already "climate refugees" are on the move, after floods drowned crops in recent years. Some pile into the slums ringing Dhaka; about half a million arrive each year, 70 percent of them for environmental reasons.[126] Already neighboring India is worried they won't stop there; "if one-third of Bangladesh is flooded, India can soak in some refugees, but not all," the former commander of the country's air force warned recently.[127] Worried enough, in fact, that India has spent the last five years quietly building a 2,500-mile-long wall, modeled on the West Bank barrier between the Palestinians and Israelis, portions of which will be electrified.[128] This may work for a while, but as one military historian put it, "People will pay no attention to borders. They will swamp borders. They will trample over them in desperation."[129] But even if they make it to India, they may find their troubles following close behind. The closest big Indian city is Calcutta, where a new report found rising seas pushing saltwater up to its borders; mangrove trees, normally found a hundred kilometers downstream along the Ganges, were beginning to colonize the city's riverbanks. "We have already spotted more saline water fish in the river," one official reported.[130]

The U.S. military—besides worrying about sixty-three of its own coastal facilities that are in danger of flooding as sea level rises—has begun planning for a future where "climate change will require mass mobilizations of the military to cope with humanitarian disasters," in the words of one researcher. The Defense Department has to ask "if our forces are adequate enough to respond to several more Katrinas," said retired U.S. Army general Paul Kern. We used to worry about having to fight two wars at once—now the military worries about "a string of bad events, of landslides, tornadoes, and hurricanes."[131] In August

2009, the Defense Department reported that "recent war games and intelligence studies" could "destabilize entire regions." The scenarios "get real complicated real quickly," one deputy assistant secretary of defense explained.[132] The most lurid account of all came from a Pentagon-sponsored report forecasting possible scenarios a decade or two away, when the pressures of climate change have become "irresistible—history shows that whenever humans have faced a choice between starving or raiding, they raid. Imagine Eastern European countries, struggling to feed their populations, invading Russia—which is weakened by a population that is already in decline—for access to its minerals and energy supplies. Or picture Japan eyeing nearby Russian oil and gas reserves to power desalination plants and energy-intensive farming. Envision nuclear-armed Pakistan, India, and China skirmishing at their borders over refugees, access to shared rivers, and arable land." Here's the bottom line from that Pentagon report, a picture of a new planet that, at least as far as conflict goes, resembles nothing so much as the old one: "Wars over resources were the norm until about three centuries ago. When such conflicts broke out, 25% of a population's adult males usually died. As abrupt climate change hits home, warfare may again come to define human life."[133]

Well, that's a tad grim. Not really the career I trained for, fighting other adult males over the fall harvest. And I don't think we need to go there. The second half of this book is based on the premise that we can build durable and even relatively graceful ways to inhabit this new planet.

But first we really do need to come to terms with where we are. We need to dampen our intuitive sense that the future will resemble the past, and our standard-issue optimism that the future will be ever easier. *We do not live any longer on the flat earth that Tom Friedman postulated.* Eaarth is an uphill planet now, where gravity exerts a stronger pull than we're used to. There's more friction than we're used to. You have to work harder to get where you're going.

I think we know that in our bones—I think we felt it even before the Bush recession settled over us. For Americans, the crucial moment may have come earlier in 2008, six months before the big banks started tottering, at the moment when the economy still seemed to be roaring, but the cost of gasoline suddenly spiked to four dollars a gallon. If the American idea has one constant, it's motion. We arrived here from distant shores; we crossed the continent; we built the highway; we invented the GPS box that sits on your dashboard telling you that you missed your turn. Everything was moving right along. And then, all of a sudden, really for the first time, that motion began to lurch. Each month Americans drove less than the month before. You couldn't sell your old house—but you *really* couldn't sell your old Explorer. Suddenly, in fact, you felt a little less confident that you were an Explorer, a Navigator, a Forester, a Mountaineer, a Scout, a Tracker, a Trooper, a Wrangler, a Pathfinder, a Trailblazer. You all of a sudden *were* in Kansas or maybe in New Rochelle—not Durango, or Tahoe, or Denali, or the Yukon. Discovery and Escape and Excursion suddenly seemed less important than the buzz-killing fact that it took a hundred bucks to fill the tank. In Japan, after its lost decade of economic floundering, consumer researchers reported that the percentage of people wishing to own a car had dropped by half.[134]

Or consider flight, the quintessential twentieth-century American invention. If the world seems flat to people like Friedman, it may be because they spend so much time up above it; get high enough and things smooth out. Until recently everyone agreed that each year would bring more fliers—the FAA predicted domestic traffic would double by 2025, to 1.3 billion passengers. But when the price of oil soared, planes landed. Thirty cities lost commercial air service in 2008—cities like Wilmington, Delaware. In 2009, British Airways asked its employees to work a month for free to help it stay airborne.[135] Two transportation experts, Anthony Perl and and Richard Gilbert, recently published a book envisioning a future in which rising oil prices cut American air traffic 40 percent by 2025 and the number of primary airports falls from four hundred to fifty. How is the air force preparing for peak oil? With a plant that will transform coal to jet fuel, the single most effective strategy for warming the planet anyone has yet discovered.[136] The oil analyst Richard Heinberg describes giving the keynote address at the world's first Electric Aircraft Symposium. "NASA and Boeing sent representatives, but all told there were about 20 in attendance. The planes being discussed were ultralight two-seaters: that's the limit of current or foreseeable battery technology."[137]

If aviation defined the twentieth century, endlessly expanding trade has been the mark of our new era. Instead of addressing climate change, Bill Clinton invested his political capital in passing the North American Free Trade Agreement and the General Agreement on Tariffs and Trade, opening first the continent and then most of the planet to essentially unrestricted commerce. To say that these agreements "worked" is an understatement: by 2005 the Koreans had built a container ship large enough to carry eight thousand of the twenty-foot units—lined up end to end, they would stretch for fifty-three kilometers.[138] A year later, the Chinese topped that with a ship big enough to haul 1.3 million twenty-nine-inch color TVs, or 50 million mobile phones. (It was named the *Los Angeles*, for the port where it dropped off much of its cargo, before returning empty.)[139] A few months later, a Danish ship half again as large made its maiden voyage, dubbed the *SS Santa* for the cheap toys it carried from Shanghai to the United Kingdom. It returned home laden with British waste destined for Chinese recycling centers.[140] In 2005, 7 million storage containers were off-loaded in the United States, and only 2.5 million went back to Asia with anything aboard; architects started building starter housing with the steel boxes.[141] Before long the American trade deficit had ballooned, and a couple of hundred million Chinese had made their way off the farm to the factory, raising their income overnight—and raising a cloud of carbon into the atmosphere. All in all, the global volume of trade grew *fivefold* in the quarter century that began with Ronald Reagan's inauguration.[142]

But that was on the old planet. In 2008, something odd started happening. First, as the price of oil spiked, shipping things long distance started to seem less attractive. By May, the cost of sending a shipping container from Shanghai to the United States was eight thousand dollars, up from three thousand dollars at the beginning of the decade. Cargo volumes began to fall—Ikea opened a plant in Virginia, not China. "The low-hanging fruit of globalization has been picked," a Morgan Stanley currency strategist said. Jeff Rubin, an analyst with CIBC World Markets in Toronto, was blunter: "Globalization is reversible."[143] Indeed, Midwest steelmakers reported a surge in demand, precisely because "soaring transport costs, first on importing iron to China and then exporting finished steel overseas, have already more than eroded the wage advantage and suddenly rendered Chinese-made steel uncompetitive in the US market."[144] A poll of U.S. and European manufacturing executives in late 2008 found 90 percent "contemplating bringing some operations home or at least closer."[145]

As the price of oil rose, and with it the demand for ethanol, the cost of food also soared—and suddenly nations began deciding that free trade was not as blindingly obvious as they'd once insisted. The "Doha Round" of world trade liberalization talks collapsed in midsummer 2008, after "growing worries in China about food security" overrode "the country's previous commitment to free trade." The Chinese joined the Indians, who had long worried that Western agriculture would overwhelm their own farming economy. India's top trade negotiator insisted that "every country must first ensure its own food security."[146] "It is a massive blow to confidence in the global economy," a spokesman for the European Commission said;[147] in an editorial the *Wall Street Journal* wondered if it marked "the end of the post-World War II free-trade era."[148] When the financial panic set in a month or two later, trade dropped off sharply; global shipping costs plunged 90 percent, and suddenly you could rent an enormous bulk carrier for $5,600 a day, down from $234,000 a few months before.[149] Even so, the joke in Singapore, once one of the globe's busiest ports, was that so many idle vessels waited in the harbor that you could walk across them to reach Indonesia.[150] There were also reports from American harbormasters that "from southern California to Maine, high fuel prices and poor fishing have driven boat owners to abandon perhaps thousands of vessels on the waterfront, where they are beginning to break up and sink."[151] Surveying the wreckage, one investment analyst said simply that globalization was "the biggest bubble of them all."[152]

There's something viscerally disturbing about idled ships. Eventually some of the great vessels returned to sea, as economies began slowly to improve, but trade volume for all of 2009 was down for the first time in more than a quarter century. And some of the trade was very different. The governments of Thailand and Iran, for instance, worked out a direct barter operation, rice for oil, "a system of trade not used for decades."[153]

It's possible that just as we've seen the peak of oil, we've also seen the peak of economic growth—that we won't be able to make the system bigger. If not now, then soon. Insurance costs climb, the price of oil spikes, the economy tanks, the money for new investment in energy evaporates, and when the economy starts to accelerate again the price of oil spikes. In May 2009, a study from McKinsey and Company called a new oil shock "inevitable."[154] Lather, rinse, repeat. Except that the temperature keeps rising. So lather, and then stand here with a head full of suds because the rising temperature has evaporated your reservoir.

Of all the things I've told you about our new planet—the stronger storms, the melting ice, the acid ocean—the most terrifying and strangest change would be the end of growth. Growth is what we do. Who ever really dreamed it might come to an end?

Who indeed. Back in a very different time, when Lyndon Baines Johnson was president, in the spring of the My Lai massacre and Dr. King's assassination and the Broadway opening of *Hair*, a small group of European industrialists and scientists met in a villa in the Italian capital. Their group—the Club of Rome—proposed to examine interrelated global trends, and they commissioned a report from a team of young systems analysts at the Massachusetts Institute of Technology.

By the time that team had finished its work and issued it as a book called *Limits to Growth* in 1972, the first Earth Day had taken place, and Richard Nixon had created the Environmental Protection Agency. But few events in environmental history were more significant than the publication of that slim book, which was translated into thirty languages and sold 30 million copies, more than any other volume on the environment. In certain ways its concerns seem almost quaint and musty to us now: careful tables of the concentration of DDT in the body fat of Eskimos and Germans and Israelis, or earnest talk about *pollution,* a word we hardly hear anymore. But more clearly than anyone else, this small team of researchers glimpsed the likelihood that we would overwhelm the planet on which we lived, and in so doing make our lives much harder. They foresaw this planet Eaarth, and if we'd heeded them we might have prevented its birth.

So it's worth, if only for history's sake, looking back at their work. They charted the exponential growth of several kinds of emissions, and they also calculated the depletion of a variety of resources, especially fossil fuel. Plugging these and other variables into their computer, they produced a series of model runs that showed much the same thing: humanity was very likely to "overshoot" and then collapse. In some of the models, "industrial capital stock grows to a level that requires an enormous input of resources. In the very process of that growth it depletes a large fraction of the resource reserves available. As resource prices rise and mines are depleted, more and more capital must be used for obtaining resources, leaving less to be invested for future growth." This sounds marvelously like the rise of China and the onset of peak oil, though they could not have named the country or the resource with the data then available. In scenario after scenario they tweaked the model to allow for more minerals in the ground, or more atmosphere to absorb pollution, but they kept finding that the momentum of economic growth was so large it overwhelmed these fixes. They concluded three things:

1. If the present growth trends in world population, industrialization, pollution, food production, and resource depletion

continue unchanged, the limits to growth on this planet will be reached sometime within the next 100 years.

2. It is possible to alter these growth trends and to establish a condition of ecological and economic stability that is sustainable far into the future. The state of global equilibrium could be designed so that the basic material needs of each person on earth are satisfied and each person has an equal opportunity to realize his or her individual human potential.
3. If the world's people decide to strive for this second outcome rather than the first, the sooner they begin working to attain it, the greater will be their chances of success.[155]

The book was wonkish in the extreme—page after page with graphs of different model runs, the lines all soaring and then crashing. That wonkishness was part of its appeal: this was the first time "a computer" had weighed in on the fate of humanity. It seemed so scientific; the book inspired headlines like "A Computer Looks Ahead and Shudders."[156] Given the moment, there were millions of fervent supporters; given the stakes, there were also, instantly, fierce critics. They presented the obvious objections: new technology will overcome any problems; if the price of oil goes up, we'll find more of it; you can't really model the world on a computer; people who say otherwise are commie hippies. One book of responses published a year after the original concluded with the charge that "like the great prophet of world salvation through world breakdown, Karl Marx, their apocalyptic visions of the immediate future are tempered by the glittering image of utopia barely discernible through the fire and brimstone that rages in the historical background."[157]

What's amazing, in retrospect, is how close we came to actually listening to their message. Around the world, people got to work figuring out how to slow population growth. Educating women turned out to be the best strategy, and so we've watched the average mother go from having more than six children to fewer than three in amazingly short order. We were paying attention; these were the years of the first oil crises, the first big tanker spills, the first fuel economy standards for cars. Heck, these were the years when America adopted the fifty-five-mile-per-hour speed limit—when we actually slowed down our mobility in the name of conservation. E. F. Schumacher published his opus, *Small Is Beautiful,* two years after *Limits to Growth,* and it too became a best seller; President Jimmy Carter hosted a reception for Schumacher at the White House. That is, there was a White House gathering for a man advocating "Buddhist economics," whose only critique of the MIT team was that they had not gone far enough, a man who insisted that right now, while the world was still working properly, we needed to take dramatic action—needed to build a "lifestyle designed for permanence," to embrace "the evolution of small-scale technology, relatively nonviolent technology, technology with a human face."[158] That's the guy Carter had over for dinner. The president also, famously, stuck solar panels on the White House roof; he even went on television wearing a sweater. Turn down the thermostat, put on the cardigan! At the time, such gestures resonated. The sociologist Amitai Etzioni reported to Carter that 30 percent of Americans were "pro-growth," 31 percent were "anti-growth," and 39 percent were "highly uncertain."[159] Got that? In the late 1970s more Americans were opposed to continued economic growth than in favor, something that seems almost impossible to us now. But it meant we actually had a brief opening to steer a different course, away from the rocks.

We didn't, of course. Etzioni told Carter that such ambivalence was "too stressful for societies to endure," and predicted we'd make a decision one way or the other. That would come in the election of 1980. I remember that election well. I was covering the contest for my college newspaper and spent weeks on Reagan's press bus; I even got a sit-down interview with him. (He couldn't have been more gracious; Nancy couldn't have looked at her watch less than twenty times.) And on that November night I wrote the story of his triumph and then got drunker than I've ever got in my life, a despairing drunk.

When I eventually woke up, I wrote a long essay, as jejune as most of what I wrote in those days. But somehow I understood at least a little of what it meant: that we'd passed up our last good chance to brake our momentum and head in a new direction. We were trying to expand our way out of Carter's economic trouble, I wrote, "in an age when the fact of scarcity—the idea Reagan and his men hate most, that they try to hex away with statistics about oil reserves and mineral supplies—makes even the myth of an ever-expanding industrial economy seem ludicrous." But I knew that they'd get away with it—that there was still enough slack in the system that it wouldn't come crashing down on *them*. Deregulation—the relentless notion that government was the problem and that freed from its strictures the economy would grow—was in the short term remarkably effective. We would be able to flog the economy back to life, give it a dose of economic amphetamine that would last ten or fifteen years, "the years when our slide will gain invisible but irresistible momentum."[160] The problem was not that Reagan's sunny optimism somehow masked a fascist soul; the problem *was* his sunny optimism. He really believed it was morning again, and when the economy turned up, so did the rest of the country; the ambivalence about growth vanished, and with it our last real chance to avert disaster. Because the next twenty-five years, all lived in Reagan's shadow, were the years when we pumped the carbon dioxide into the atmosphere and the oil out of the ground.

In the short run, Reagan took the solar panels off the White House roof, and he froze the mileage standards that had helped cut oil demand by more than a sixth in a decade.[161] (Long before he left office, we stopped driving fifty-five.) In the slightly longer run, his worldview gave us not only the Bush administrations but also the Clinton years, with their single-minded focus on economic expansion. The change was not just technological; it wasn't simply that we stopped investing in solar energy and let renewables languish. It's that we repudiated the idea of limits altogether—we laughed at the idea that there might be limits to growth. Again, not just right-wing Republicans but everyone. Here's Larry Summers, treasury secretary under President Clinton, now Obama's chief economic adviser: "There are no . . . limits to the carrying capacity of the earth that are likely to bind any time in the foreseeable future. There isn't a risk of an apocalypse due to global warming or anything else. *The idea that we should put limits on growth because of some natural limit is a profound error.*"[162] And here's a taunting ad, circa 2002, from Exxon Mobil, which funded the fight against action on global warming: "In 1972, the Club of Rome published 'Limits to Growth,' questioning the sustainability of economic and population growth. . . . The Club of Rome was wrong."[163]

Not wrong, as it turned out. Just ahead of the curve. You can ignore environmental problems for a long time, but when they catch up to you, they catch up fast. I remember visiting the Canadian province of Newfoundland in the early 1990s, not long after authorities imposed an all-out ban on fishing for cod, the activity that had defined the island since Europeans arrived there five hundred years ago. In fact, the ban was hardly necessary, because in 1992 fishing boats had gone out and found, basically, not a fish. What happened? Fish harvests had been steady in the years before 1992, but not, it turned out, because stocks were healthy. Instead, "there were new advances in fish-finding technology every couple of months," explained a professor who studied the collapse. New sonar, new computer gear—what fish remained were rooted out. Nobody realized that they were scraping the bottom of the barrel until it literally came up empty. Now, the hundreds of outports that ringed the coast of Newfoundland are dead or dying—I met fishermen who were being "retrained" by the government to style hair.[164]

On a much larger scale, that's what happened with the encroaching limits to growth. We fixed a few relatively easy environmental problems, like smog, or filth leaking into our rivers, and that was enough to convince us that we were making progress. Lake Erie wasn't burning anymore. But all the while the sheer growth in the size of our enterprise was making those old predictions from the 1970s come steadily, invisibly true. Global per capita production of grain peaked in the 1980s. Total global fish catch peaked in the 1990s. Fifty-four nations—mostly those with real environmental troubles—saw their per capita GNPs decline during the 1990s even while we were booming.[165] In June 2008, an Australian academic, Graham Turner, looked at every system that the original *Limits to Growth* had actually examined—population, agriculture, industry, pollution, and resource consumption—and found that "for the first thirty years of the model, the world has been tracking along the unsustainable trajectory of the book's business-as-usual scenario." The curves, he said, matched the standard model, the one that ended in economic collapse sometime before midcentury. "The contemporary issues of peak oil, climate change, and food and water security resonate strongly with the overshoot and collapse displayed" in the *Limits to Growth* model, he said—adding, with the always-a-silver-lining attitude of a true academic, "It's been a rare opportunity to evaluate the output of a global model against observed and independent data."[166]

Basically, it turns out they were right. The Club of Rome, E. F. Schumacher, the MIT crew with the whiz-bang program on their 1970 mainframe, Jimmy Carter. They were right. You grow too large, and then you run out of oil and the Arctic melts.

I've belabored the point. I've belabored it because by now every force in our society is trained to want more growth. If we get in a fix, people like Tom Friedman immediately tell us that we need to get bigger, but in a green way. Because the world is "a growth machine" that "no one can turn off." Look, it's easy to make sour fun of this. Friedman's wife, with whom he built an eleven-thousand-square-foot mansion near Washington, is the heir to something with the wonderfully apt name of General Growth Properties. The company operates malls across the United States—in Farmington, New Mexico, for instance, "where ancient Anasazi traders came to its three rivers to exchange goods and services" and now the "Animas Valley Mall continues that longstanding tradition" with a Dillards and a Fuddruckers. Also, the Grand Canal Shoppes at the Venetian in Las Vegas. The Anasazi? Venice? Like I say, easy to make fun of. But there's plenty of reason to think the way he thinks. The way our economy works at present, any cessation of growth equals misery. If foot traffic at Fuddruckers ebbs, the cook can't get health care. And if we can't grow, we can't easily pay off the massive debts we've incurred. It *would* be nicer to fire up the engines one more time, à la Reagan. But we *can't* grow. General Growth Properties saw its stock fall from fifty-one to thirty-five cents a share in the great crunch. Then it went bankrupt.[167] There's too much friction. We're on an uphill planet. So we'd better change.

• • •

There's been a certain fascination with Easter Island in recent years, and with the Greenland Norse, and with the other stars of the new genre of what you might call "collapse porn." From Jared Diamond's superb *Collapse* to Jim Kunstler's dark and funny novel *World Made by Hand*, a score of books have given us the slightly scary shiver of imagining our lives tumbling over a cliff. As one English newspaperman put it, "The Maya, like us, were at the apex of their power when things began to unravel. . . . As stock markets zigzag into uncharted territory and ice caps continue to melt, it is a view increasingly echoed by scholars and commentators."[168] (For some reason it makes me giggle to imagine the Mayan Bernie Madoff.) The *New Yorker* ran a feature on "the new dystopians"—"doomers," it called them—people advising that you buy pistols or hoard gold or corner the market on firewood.

It's not that such collapse is impossible, not by any means. All the charts in *Limits to Growth* seem to show it's nearly inevitable, since we didn't take our foot off the gas when we had the chance. And the seeming solidity of our civilization could be illusory. As Diamond shows with abundant example, "one of the main lessons to be learned from the collapses of the Maya, Anasazi, Easter Islanders, and those other past societies is that a society's steep decline may begin only a decade or two after the society reaches its peak numbers, wealth and power." (In fact, that's exactly when you'd expect it to happen, because that peak wealth usually means peak impact on the environment.)[169] We're overleveraged in ways that resemble those past civilizations. In fact, they were pikers compared to us: the Anasazi may have died out when the climate shifted, but they didn't *make it happen*. And if our societies start to tank, we'll be in worse shape than those who came before. For one thing, our crisis is global, so there's no place to flee. For another, most of us don't know how to do very much—in your standard collapse scenario, it's nice to know how to grow wheat.

The trouble with obsessing over collapse, though, is that it keeps you from considering other possibilities. Either you've got your fingers stuck firmly in your ears, or you're down in the basement oiling your guns. There's no real room for creative thinking. To its theologians, collapse is as automatic and involuntary as growth has been to its acolytes.

The rest of this book will be devoted to another possibility—that *we might choose instead to try to manage our descent.* That we might aim for a *relatively graceful decline.* That instead of trying to fly the plane higher when the engines start to fail, or just letting it crash into the nearest block of apartments, we might start looking around for a smooth stretch of river to put it down in. Forget John Glenn; Sully Sullenberger, ditching his US Airways flight in the Hudson in January 2009, is the kind of hero we need (and so much the better that he turned out to be quiet and self-effacing). Yes, we've foreclosed lots of options; as the founder of the Club of Rome put it, "The future is no longer what it was thought to be, or what it might have been if humans had known how to use their brains and their opportunities more effectively." But we're not entirely out of possibilities. Like someone lost in the woods, we need to stop running, sit down, see what's in our pockets that might be of use, and start figuring out what steps to take.

Number one is: mature. We've spent two hundred years hooked on growth, and it's done us some good, and it's done us some bad, but mostly it's gotten deep inside us, kept us perpetually adolescent. Americans in particular: Edward Everett, the governor of Massachusetts, gave a speech in 1840 in which he said, "The progress which has been made in art and science is, indeed, vast. We are ready to think that the goal must be at hand. But there is no goal; and there can be no pause; for art and science are, in themselves, progressive and infinite. Nothing can arrest them which does not plunge the entire order of society into barbarism."[170] In the vernacular of our time, here's the economics columnist Robert Samuelson, writing in *Newsweek*: "We Americans are progress junkies. We think that today should be better than yesterday and that tomorrow should be better than today."[171] Every politician who ever lived has said, "Our best days are ahead of us." *But they aren't*, not in the way we're used to reckoning "best." On a finite planet that was going to happen someday; it's just our luck that the music stopped while we were on the floor. Yes, it's tough—but then, it's been tough for other people in other times and places. So if 2008 turned out to be the year that growth came to an end—or maybe it will be 2011, or 2014, or 2024—well, that's the breaks. Harder for the Chinese than for us; they'd just begun to taste some of that ease. Or maybe easier for them, since they're less used to it. But it is what it is. We need to see clearly. No illusions, no fantasies, no melodrama.

That's easier said than done—we all want to hold on to the vague idea that we can make it work. If you're in the developed world, that might mean embracing "geoengineering" schemes: filling the atmosphere with sulfur to block sunlight (on-purpose smog), or filling the seas with iron filings to stimulate the growth of plankton that would soak up carbon. But the early tests have found only "negligible" results,[172] and the costs are huge, measured in the tens of trillions of dollars.[173] Not only that, but we'd be experimenting on the same scale that we've experimented with carbon, and look how well that's turned out. I have more sympathy with the daydreams of the developing world. At a recent meeting of Asian journalists, for example, one delegate suggested that Bangladesh could be relocated to Siberia and Iceland, because melting snows would turn them into "breadbaskets."[174] How to tell them instead that the tundra is turning into a methane-leaking swamp?

Step number two: we need to figure out what we must jettison. Many habits, obviously—little things like the consumer lifestyle. But the big item on the list becomes increasingly clear. *Complexity* is the mark of our age, but that complexity rests on the cheap fossil fuel and the stable climate that underwrote huge surpluses of food. With that cushion, we were able, in Richard Heinberg's words, "to elevate social complexity to an art form." Unlike other animals who "get up in the morning and simply start milling around looking for food," we "get up in the morning and . . . well, here the story diverges in millions of ways. Some of us commute to offices or factories. Some people have jobs building or maintaining the cars we drive. Other people have jobs reading the news we listen to on the radio as we navigate the freeway."[175] That complexity is our glory, but also our vulnerability. As we began to sense with the spike in oil prices and then the credit crunch in 2008, we've connected things so tightly to each other that small failures in one place vibrate throughout the entire system. If America's dumb decision to use a fraction of its corn crop for ethanol can help set off food riots in thirty-seven countries, or if a series of shortsighted bets on Nevada mortgages can double unemployment in China, we've let our systems intertwine too much. If our driving habits can move the monsoon off the Asian subcontinent or melt the Arctic ice cap—well, you get it.

We've turned our sweet planet into Eaarth, which is not as nice. We're moving quickly from a world where we push nature around to a world where nature pushes back—and with far more power. But we've still got to live on that world, so we better start figuring out how.

· 3 ·

BACKING OFF

We recoil when faced with a future different from the one we imagine. And it's hard to brace ourselves for the jump to a new world when we still, kind of, live in the old one. So we tell ourselves that the scientists may be overstating our environmental woes, or that because our stock market has climbed back from its lows we'll soon be back to the old growth economy. As we've seen, though, scientists are far more guilty of understatement than exaggeration, and our economic troubles are intersecting with our ecological ones in ways that put us hard up against the limits to growth. This book has been dedicated, so far, to the idea that we're in very deep trouble. Now we must try to figure out how to survive what's coming at us. And that survival begins with words.

We lack the vocabulary and the metaphors we need for life on a different scale. We're so used to *growth* that we can't imagine alternatives; at best we embrace the squishy *sustainable,* with its implied claim that we can keep on as before. So here are my candidates for words that may help us think usefully about the future.

Durable
Sturdy
Stable
Hardy
Robust

These are squat, solid, stout words. They conjure a world where we no longer grow by leaps and bounds, but where we *hunker down*, where we *dig in*. They are words that we associate with maturity, not youth; with steadiness, not flash. They aren't exciting, but they are comforting—think husband, not boyfriend.

Here's a better metaphor: the economy that has defined our Western world is like a racehorse, fleet and showy. It's bred for speed, with narrow, tapered legs; tap it on the haunch, and it accelerates down the backstretch. But don't put it on a track where the rain has turned things muddy; know that even a small bump in its path will break its stride and quite likely snap that thin and speedy leg. The thoroughbred, like our economy, has been optimized for one thing only: pure burning swiftness. (Also, both are now mostly owned by sheikhs.) What we need to do, even while we're in the saddle, is transform our racehorse into a workhorse—into something dependable, even-tempered, long-lasting, uncomplaining. Won't go fast, will go long; won't win the laurel, will carry the day. The high praise for a workhorse—for a Shire or a Belgian or a Percheron—is "she's steady." "She can pull." We're talking walk or trot or jog, not canter or gallop.

Our time has been marked by ever-increasing speed—paddle-wheeler to locomotive to airplane to rocket, Model T to Formula 1. Can you imagine slower? Maybe so: the Slow Food movement has spread steadily around the world for a decade. Now there's Slow Design, embracing the return of handwork;

and the Slow City campaign. Our time has been marked by great ups and downs, booms with the occasional bust. Can you imagine steadiness? Can you make it work in your mind?

Most of all, of course, our time has been the time of bigness—the amazing ever-steepening upward curve, where things grew and grew and then grew some more. Economies and road networks and houses, inflating until there were entire subdivisions filled with starter castles for entry-level monarchs. Stomachs and breasts and lips, cars and debts, portions and bonuses. *Can we imagine smaller?* That is the test of our time, both practical and psychological. How can we adjust to the fact that we're not going to get bigger? That the wind has begun to blow harder, and hence we need to lower our wind resistance. That the sun has begun to burn more brightly, and hence like other animals we need to reduce our size. That the oil has begun to run out, and hence everything we own is suddenly larger than it should be.

It's not easy to make the imaginative leap. Slow food is one thing, but *shrinking* is another. The pain of the *recession*—a word that, after all, literally means getting smaller—has been entirely real, because our economy is geared to work only with growth. By definition, the only way to escape recession is to grow larger again.

On the other hand, in certain ways our economic trouble gives us real insight into scale, makes it easier to start thinking subversive thoughts. If there is a phrase that sticks in the mind (and in the craw) from the last few years, it is "too big to fail." Giants like Citibank or AIG had swelled to a size where if they collapsed they could bring down the entire financial system. The phrase, loosely translated, meant "the government must bail us out." That's the part we debated on TV and in the op-ed columns: should we prop them up? But the simpler meaning of the phrase was that they were "too big." Anything too big to fail is by *definition* too big.

We thought we'd spun a kind of magic that would suspend the laws of gravity; ever since Reagan, the libertarian economists had insisted that self-interest alone was enough to hold at bay the possibility of collapse. That's why the amount of retail space per person in the United States doubled, from nineteen to thirty-eight square feet, between 1990 and 2005.[1] It didn't make rational sense, but it made sense as long as the magic held. The spell broke late in the summer of 2008, and after that there was only poor Alan Greenspan looking less like the master magician and more like the tired, tiny wizard behind the curtain. His belief system had turned out to be "flawed," he testified to Congress. "I made a mistake in presuming that the self-interests of organizations, specifically banks and others, were such as that they were best capable of protecting their own shareholders and their equity," he said—and you almost felt sorry for him. I mean, we'd lost our money, but he'd lost an entire belief system. As he put it, "The whole intellectual edifice collapsed in the summer of last year because the data inputted into the risk management models generally covered only the last two decades, a period of euphoria."[2]

That's actually a sentence worth rereading. On a larger scale, our whole civilization stands on the edge of collapse because the data inputted into our risk management models come from the last couple of hundred years, a very atypical time. A giddy time, high on oil. It's not just the banks that have gotten too big to fail, but all the arrangements of modern life. Our time, on every front, has been marked by the dizzying Alice-on-her-first-pill explosion in the size of the human enterprise. For almost all of human history, our society was small and nature was large; in a few brief decades that key ratio has reversed. Most of the time it's happened just a little too slowly for us to really feel it, but every once in a while there's been a flash. The first to announce it was probably J. Robert Oppenheimer, watching the first nuclear explosion at Alamogordo in the New Mexico desert. He quoted from the Bhagavad Gita: "We are become as Gods, destroyers of worlds." That particular kind of explosion is easy enough for us to imagine (especially after Hiroshima and Nagasaki); hence we have so far done what we can to hold it at bay. It is far harder for us to imagine that the explosion of a billion pistons every minute can be doing damage on the same kind of scale. But that's us. Big.

Why, after a certain point, does bigness spell trouble?

For one thing, useful feedback diminishes as scale expands—you're too far away from reality. Perhaps the finest early piece of reporting on the financial crisis came from the radio program *This American Life*, which in May 2008 aired an episode called "Giant Pool of Money." The reporter Alex Blumberg meticulously traced home mortgage loans up and down the chain of failure from the strip mall operations across the Sun Belt where they began. He interviews a loan officer in Reno who describes handing out half-million-dollar loans to people with no hope of repaying them. He finds a guy in upstate New York working out of a giant boiler room who was clearing $100,000 a month originating these kinds of loans and immediately selling them off to the big banks and brokerages that the taxpayers later bailed out. "These people couldn't make a car payment and we're giving them a $400,000 house,"[3] the New Yorker says. The local guys wouldn't have made the loan if they'd had to collect on it themselves, and Citibank wouldn't have bought the loan if it had ever met the deadbeat on the other end, but because the system was so big the crucial pieces of information never got where they should have. And of course no one wanted to know any more than they needed to, because for a while they were all profiting so sweetly.

Contrast that with, say, the First National Bank of Orwell. Orwell is a small Vermont town, down the mountain from my house. In November 2008, at the very height of the panic, a reporter from the *New York Times* visited and found the bankers still making loans, which were still being repaid. "We're very particular on what and who we take," said Bryan Young, the bank's vice president. The bank still has brass cages for the tellers, but that's because it's frugal, not sentimental. Not sentimental at all. "Having the borrower sit in front of us is very meaningful. We're not massive brokers, trying to underwrite a loan from the 55th floor of an office somewhere. There's serious value in looking someone in the eye and understanding what their drive is, where they're coming from, and how serious they are about the project."[4] Value, too, in being able to drive out and look at the site where someone is building something with your money. The bank in Orwell is not unique. Even as the financial crisis began to worsen, the FDIC reported that most smaller institutions were not in any trouble. It wasn't just bucolic rural banks. Broadway Federal, in Los Angeles, was founded in the 1940s to serve the city's black neighborhoods—five branches, "copious concrete, fluorescent lights, clunky logo." In their story on the bank, Phillip Longman and T. A. Frank described the feedback loops for a small institution as "informational capital"—it's probably the biggest reason that failure rates are seven times greater among banks with assets above a billion dollars.[5]

Eventually, of course, recession spreads even to these smaller, smarter banks. The losses from irresponsible mortgages damage the economy, which in turn costs responsible borrowers their jobs, and so on. But the crucial fact is that the damage can't spread the other way. There's nothing that Broadway Federal could do to trigger a recession, and that's the other advantage of smallness: mistakes are mistakes, not crises, until they're interconnected into a massive system. I have solar panels on my roof, and if something happens to them (so far nothing has—they've got very few moving parts), I have a problem. But it doesn't cause a problem for my neighbors, the way it would if a giant power plant or two suddenly crashed. Terrorism is not an issue; even if a terrorist made his way to my rooftop with a hammer and smashed my panels, they would not spew deadly solar particles into the atmosphere. Many small things breed a kind of stability; a few big things endanger it—better the Fortune 500,000 than the Fortune 500 (unless you want to be an eight-figure CEO).

In an easy world—on the kind of planet with plenty of cheap fossil fuel and a stable climate—bigness has some advantages. But we don't live on that kind of planet anymore. At the end of the Cretaceous period, when an asteroid strike quickly changed the climate and apparently brought the long-running reign of the dinosaurs to an end, the mammals that made it through the disaster were rat-sized or smaller because, as a team of geologists put it in 2004, "being too large to find a hole to hide in would have been a death sentence."[6] Small doesn't automatically save you; lots of little birds and reptiles died off at the end of the Cretaceous, too. In 2008, little Iceland managed to crash its economy—by behaving big, by turning its banks into casinos, a purposeful effort to become big big big. "It wasn't corruption that did in the Icelandic economy," wrote the journalist Rebecca Solnit in one of her dispatches from the crisis. "It was government-led recklessness and deregulation, against the backdrop of a passive population fixated on the new wealth." Now that's all gone, and they're starting over—going back to fishing, for instance, long the only source of real wealth on the island. But even in Iceland, at the height of the financial crisis, small communities were doing better; they've "neither been affected by the current economic crisis nor were they included in the period of expansion before the crisis arrived," Solnit reported.[7] In Lagansbyggd, the mayor said, "We need a nurse. And we need workers to fill jobs both at sea and on land."[8]

If we're talking big and small, there's no avoiding the biggest object of them all: the United States itself. I grew up in Lexington, Massachusetts, which in most ways is a typical American suburb—excellent and progressive school system, overpriced houses, not as much to do as a teenager might wish. But of course it's also the "Birthplace of American History," the site of the first battle of the American Revolution and hence of any revolution against modern empire. I spent my junior high school summers on Lexington Green, wearing a tricorne hat and guiding busloads of tourists around the Green. In those days I knew the history of the nineteenth of April 1775 in real detail. Now all I can recall are snippets: Jonathan Harrington, fatally wounded, managing to crawl across the Green so he could die bleeding on his own doorstep. Captain John Parker, who didn't actually say the words on the monument ("if they mean to have a war, let it begin here") but instead ordered his men to disperse, only to watch his cousin Jonas killed by a bayonet when fighting broke out anyway.

Telling those stories day after day was good for me; it ensured that I'd never think that patriotism and dissent were opposites. More to the point, it taught me deep down that at the start America involved the defense of the small against the big. The Minutemen weren't *Americans*, any more than Jesus was a Christian. They were in their place, next to their neighbors, defending their homes—one of them died on his *doorstep.* As one Patriots Day orator put it a century later, the British passed on from our Green "to meet the men of Acton, and Concord, and Bedford, and Carlisle, and Littleton, and Chelmsford, and Reading, and Sudbury." The "blood of Essex mingled with that of Middlesex in the great event," he insisted; "the men of Danvers lay down with the heroes of Lexington to awake to a glorious immortality."[9] That is, the Minutemen were, at the outset, defending less the idea of America than the idea of Chelmsford, of small and tight and connected communities. Of small and local economies without the margin to easily afford the various taxes and duties the British imposed. Of the idea that they should be able to figure out their own destiny—which at the time they hashed out in local town meetings each spring.

That is worth remembering now that America has come to define large—the biggest economy, the biggest military, the biggest Big Gulp soft-drinks-in-a-tub, the biggest budget and the biggest debts, the biggest ecological footprint. America's history has, from the beginning, been shaped by the debate between bigness and smallness. So far we've moved, seemingly inexorably, in one direction. But the seeds of the switch we need to make—toward something less gigantic—have been there from the beginning as well.

In fact, if continents had DNA, you would have expected North America to produce multiple sets of octuplets, not the one big bouncing baby that eventually emerged. Remember: people were willing to fight and die in the Revolution because they were jealous of their liberty. They weren't fighting on religious or ethnic grounds, nor out of dire necessity, and certainly not as vassals in the service of some feudal authority. In the weeks after Lexington, as farmers started to drift into the nascent army camp on Cambridge Common, they usually arrived in local regiments and insisted on electing their own officers; they signed up for short periods of service, and when those stretches were over they often left. George Washington came from Tidewater Virginia, a place with some sense of hierarchy. To his amazement, the new commanding general found that commissioned officers who had been barbers in private life were now shaving and cutting the hair of enlisted men.[10]

Such Americans would not easily surrender their liberties to some new distant authority. Indeed, the idea that they would come mostly to think of themselves as Americans, and not New Yorkers or Georgians, seemed slight. Since men were dying for the right not to be ruled from London, they didn't want to be ruled from some slightly less distant American capital either. Our first governing document, the Articles of Confederation, was created mostly by these radicals, and it was "a constitutional expression of the philosophy of the Declaration of Independence," in the words of one historian. There was a Congress, but it couldn't raise much money and it had sharply limited powers; "the vast field of undefined and unenumerated powers lay with the states."[11]

Even from the beginning, though, there were those Federalists who wanted a stronger central government. Washington, Hamilton, Robert Morris, Charles Carroll, and other prominent conservatives yearned for some of what was lost with British rule: "centralized government with a legal veto on state laws, the power to enact general and uniform legislation, and the power to use arms to subdue rebellious social groups within the states."[12] And though they lost the argument initially, the first decade after the Revolution showed that they had a point.

For one thing, the new states squabbled among themselves. Due to the oddities of colonial cartography, Connecticut thought it owned the Wyoming Valley in what is now Pennsylvania;

Virginia had designs on other parts of the Quaker State, and Maryland in turn on Virginia. Bigger fights erupted over commerce and trade: New York City bought its firewood from Connecticut and its dinner from what would soon be called the Garden State of New Jersey. But some in the city wanted to protect their own business against outside competition, and so tariffs were passed "obliging every Yankee sloop which came down through Hell Gate, and every Jersey market boat which was rowed across from Paulus Hook to Cortlandt Street, to pay entrance fees and obtain clearances at the customhouse, just as was done by ships from London or Hamburg." Soon the affected upstate and cross-river businessmen were holding angry protests that looked a lot like the Boston Tea Party. "Another five years would scarcely have elapsed," writes the historian John Fiske, "before shots would have been fired and seeds of perennial hatred sown on the shores that look toward Manhattan Island."[13]

Meanwhile, overseas, John Adams was trying to raise money to pay back the country's war debts, but "since the gazettes were full of the troubles of Congress and the bickerings of the states, everybody was suspicious. . . . It was only too plain, as he mournfully confessed, that American credit was dead." To raise the money, the Congress in 1781 tried to impose a 5 percent duty on imports, but of course under the Articles such a levy needed the consent of each state. The request "was the signal for a year of angry discussion. Again and again it was asked, If taxes could thus be levied by any power outside the state, why had we ever opposed the Stamp Act or the tea duties?" Massachusetts finally said yes, but Rhode Island refused and so did Virginia.[14]

And then, to really teach the lesson, came Shays's Rebellion, the 1786 uprising of debt-ridden Massachusetts farmers who tried to seize the weapons in the United States arsenal at Springfield.

They were easily defeated, but the fear of a spreading insurrection struck deep; many feared, in fact, that the British would take advantage of the internal fighting to reclaim their colonies. Shays's Rebellion shifted the mood perceptibly, something even the antifederalists realized. As the Virginian William Grayson wrote at the time, "The temper of America is changed beyond conception. I believe they were ready to swallow almost any thing."[15] Recall the mood after 9/11; suddenly liberty seemed a little less important than order.

In "The Federalist No. 10," James Madison argued for a new constitution and a stronger federal government with this pivotal claim: if you make your nation big enough, you'll water down any "faction" and prevent the whole enterprise from going off half-cocked. If you have a small enough democracy, everyone can participate, and that's *bad*: "A society consisting of a small number of citizens, who assemble and administer the government in person, can admit of no cure for the mischiefs of faction. A common passion or interest will, in almost every case, be felt by a majority of the whole." Instead, a larger republic, by definition, requires representatives from each place to travel to some central location, and in the process, by "extending the sphere . . . you take in a greater variety of parties and interests; you make it less probable that a majority of the whole will have a common motive to invade the rights of other citizens; or if such a common motive exists, it will be more difficult for all who feel it to discover their own strength."[16] For people who had been living through decades of revolutionary turbulence, the prospect of slightly less politics must have been attractive.

But here's the key thing. The Federalists could also make such a strong argument for centralized efficiency over raw democracy because *there was so much to do.* The break with England complete, there was now a job at hand, a remarkably obvious

National Project the like of which human beings had rarely if ever before encountered. *Three million Europeans were clustered at the edge of a vast and presumably wealthy continent. If it was to be settled successfully, they clearly needed a strong and unified government.* Madison, along with all the political theory, argued in "The Federalist No. 14" that if the government could just get down to work, the country would soon seem much smaller anyway: "Intercourse throughout the Union will be facilitated by new improvements. Roads will everywhere be shortened, and kept in better order; accommodations for travelers will be multiplied and meliorated."[17]

Such efforts were surely necessary. A light stagecoach, for example, might take a week to get from New York to Boston, a journey that "began at 3 o'clock in the morning. If the roads were in good condition some forty miles would be made by ten o'clock in the evening," wrote one historian. "In bad weather, when the passengers had to get down and lift the clumsy wheels out of deep ruts, the progress was much slower." There were no bridges across rivers like the Connecticut; in the winter it was relatively easy to cross on the ice, and in the summer to go by rowboat, but that left months on the shoulders of the seasons when boats were dodging ice floes. No wonder, then, "the different parts of the country knew very little about each other and local prejudices were intense." And no wonder that George Washington, before he became our first president, spent much of his time trying to build canals and turnpikes. He ran—without pay—a company chartered by the Virginia legislature to figure out how to extend navigation on the Potomac west across the mountains. "Let us bind these people to us by a chain that can never be broken," he said.[18]

So the Federalists carried the day, and indeed the next couple of years. A great swell of nationalist sentiment marked the first few years after ratification of the Constitution, aided in no small part by General Washington's enormous popularity. But the argument between big and small hadn't died out, and by 1790, when Alexander Hamilton proposed the Bank of the United States, it gave birth to the first opposition political party. Headed by Thomas Jefferson, it took power in 1801, promising to swing the balance of power back toward the states. It had some success, but only some. Already the National Project was rolling (and it was given an enormous assist by Jefferson himself, when he bought the Louisiana Territory from the French and then sent Lewis and Clark out to explore the West). Jefferson had promised to drydock the navy, for instance, on the theory that having a navy was expensive and led to wars. But the Barbary pirates plagued America's expanding overseas commerce, and so, like Barack Obama two centuries later, he had to dispatch a fleet to keep the oceans open to our shipping, a move that cost so much it complicated plans for shutting down Hamilton's bank.

The arguments raged on for the first four score and seven years of the new Republic, and they often took the form of threats to secede. And those threats came not just from the South. New England grew angry with President James Madison's prosecution of the War of 1812; Massachusetts governor Caleb Strong apparently conducted secret negotiations with the British, and delegates from around the region met in Hartford in the summer of 1814. Some were bent on secession, and others insisted on a series of constitutional amendments—but Andrew Jackson's stunning victory at the Battle of New Orleans took the wind out of their sails. Still, a close call. At the start of his oratorical career, Daniel Webster argued that the federal government could not conscript soldiers: "It will be the solemn duty of the state Governments to protect their own authority over their own militia, and to interpose between their citizens and arbitrary power."[19] Two decades later, when South Carolina wanted to ignore the "Tariff of Abominations" pushed through Congress by northern manufacturers, it was Webster who reversed course. "Liberty and Union, Now and Forever," he proclaimed.

The flashpoints were always tariffs or wars or, of course, slavery—but against them, the unifying force was always this National Project, the rapid economic expansion and the settling push west that marked America in its youth. Building canals and then railroads took serious money, often more money than private investors could raise. Without the backing of a national government and the financial stability that a national bank provided, they would have come much more slowly. Taming the vast land was expensive; the phrase that recurs in a thousand newspaper editorials and legislative speeches is "internal improvements"—rivers needed to be dredged and marked with buoys, coastlines studded with lighthouses. Most people in most places didn't want to pay: the governor of South Carolina thundered against his citizens being "grievously assessed to pay for the cutting of a canal across Cape Cod," while the Virginia and Georgia legislatures passed resolutions in 1827 roundly denouncing internal improvements, protective tariffs, and taxing their citizens to "make roads and canals for the citizens of another state."[20] Such sentiments likely reflected the majority of American opinion, but the minority backing these schemes was "energetic, greedy, and influential," in the words of one historian.[21] And so the railroads were built, and the rivers dredged. The federal government was the nation's largest landowner, and it used that land to foster the kind of farming and settlement it wished to see, to finance the railroads, to build the great land-grant colleges. This kind of spending steadily lengthened the spine of our national economy—it bankrolled Manifest Destiny.[22]

Even the Civil War was, in some sense, about these questions.

Slavery—the great and unique evil of American history—was indeed the alpha and omega of the fight, but there are plenty of other letters in the alphabet. The South, largely *because* of slavery, was a static and agrarian place. (These are not, as I will argue later, necessarily evils; in fact, *static* and *agrarian* may be precisely what we now need.) This meant that the region wasn't a full participant in the National Project; the southern powers-that-be were content to let their slaves make them rich, instead of engaging in the dynamic nation building that marked the North. And indeed most of the North was willing to let the South go on holding slaves. The rub came over expanding the institution west as the nation grew, infecting the wide-open continent with a practice seen by some as immoral and others as merely backward. Many expected that, confined to the South, slavery would die out, as it already had across Europe. But the Kansas-Nebraska Act of 1854 threatened to spread the "peculiar institution," and the modern Republican Party—Abraham Lincoln's party—was born in response. The fight over Kansas marked the start of John Brown's incendiary career; it spurred the Lincoln-Douglas debates; it helped bring things to a head.

And so the war came. And the same kind of technology, efficiency, ruthless organization, and industrialism that had made the North rich also made it unbeatable. More than that, it cemented the idea that centralized government was both inevitable and virtuous. Patriotism swelled as never before—*The Star Spangled Banner* became a staple of every public gathering. Nationalism flourished, and "Union" was a cry that went well beyond abolition. In the words of Henry J. Raymond, the editor of the *New York Times*, victory for the centralized government was the key to expansion across the continent and even into Mexico. Anything else was impossible: "There is no nation in the world so ambitious of growth and power."[23] As one Boston newspaper editorialized before the war, "California wants, and lives on the hope of getting . . . a Pacific railroad. The Cotton States have thrown every obstruction in the way of this enterprise—the rest of the country is in favor of it." And indeed, even as the war was fought, the railroads kept pushing west. Indeed, it was in 1869, at Promontory Summit, Utah, four years after the surrender at Appomattox, that the golden spike was finally driven, connecting the country sea to sea.

It's not that the National Project was necessarily good. Some parts of it—the decimation of the Indians—were as evil as slavery. We cut down the greatest temperate forests in the world and drove to the edge of extinction the greatest herds of buffalo and flocks of birds the world had ever seen. The barons who built the railroads were mostly crooks, and thousands of immigrants who might as well have been slaves died in their service. The people I admire most detested much of it. "That government is best which governs not at all," scoffed the tax-dodging Henry David Thoreau, sounding like an early incarnation of Ronald Reagan.

But if it wasn't all good, it was almost certainly inevitable. The wealth represented by those tall trees and those deep midwestern soils was going to be got. And even once the actual frontier was closed, at the end of the nineteenth century, the new frontier of technological possibility opened in the twentieth. The rapid growth of our industrial power became the new National Project, and it lasted very nearly till the end of the century.

And always it justified the bigness of our nation—though that bigness was always in dispute, right up to the end. Consider Dwight Eisenhower's fight for the Interstate Highway System. He'd first crossed the country in 1919—it took sixty-two days because the roads were so bad. The idea of a great road network grew in his mind, and so did the idea of using a gasoline tax to pay for it. States had traditionally levied such taxes and wanted to keep control. As Governor Walter Kohler Jr. of Wisconsin said: "In its tax philosophy, the Federal Government has become a voracious monster, overlooking nothing in its insatiable hunger for greater revenue. We in Wisconsin are, frankly, sick to death of Federal interference in the administration of programs which should be, and have traditionally been, the responsibility of the States." Whatever, dude—the National Project was not going to stop for Walter Kohler. Here's Lieutenant General Eugene Reybold, chief of the Army Corps of Engineers and chair of the American Road Builders Council: "President Eisenhower, in a bold stroke, has blown the lid off any milktoast, piecemeal planning. . . . There is no longer room for timidity in road building plans."[24] None indeed—by the time Glenwood Canyon on I-70 was finally paved in 1992, the system was the largest public works program in world history, costing nearly half a trillion of our dollars, running from 11,158 feet at the Continental Divide in Greenwood, Colorado, to 107 feet below sea level in Baltimore's Inner Harbor Tunnel. Along its various veins you can find the world's biggest coffeepot, the world's biggest strawberry, watermelon, and artichoke, even the world's biggest rubber band ball. Overlooking I-94 near New Salem, North Dakota, you can visit Salem Sue, the world's biggest cow. She stands thirty-eight feet tall, "her reinforced fiberglass skin taut and shiny."[25] Look on my works, ye mighty, and despair.

But now it's *done*. There's really nowhere plausible left to run a highway. And not only that; almost everything else is done, too. The last real National Projects were putting a man on the moon and crushing the Soviet empire. They shared the genetic traits of such enterprises: they were wildly expensive, the money for them couldn't be privately raised, and they more or less required a central government. Vermont wasn't going to send a man to the moon; Delaware couldn't make Moscow quail. These projects pushed and expanded our vistas: we lifted men beyond the bounds of gravity; we stretched the borders of democracy. You couldn't work these wonders with Jefferson. We love to talk Jefferson—small-scale democracy, yeoman citizen, all that. But Jefferson couldn't build the highways. If you wanted to beat the Soviets and walk on the moon, you needed Hamilton and the big-money big-government schemes. And so Hamilton is what we've had.

Now, though, the list of National Projects has dwindled. Fighting Muslim terrorism turns out to require small, careful strikes, not massive weaponry. Theoretically we're committed to sending a man to Mars, but I know very few people who either believe we will or care. The only likely candidate for a new National Project is some version of Thomas Friedman's vision—the smart energy grid, an array of high-tension lines stretching out across the horizon. In fact, though—as I shall show in chapter 4—it makes more sense to think about energy locally and regionally. (The very physics of electricity, the juice lost in transmission, works against long-range strategies.) We've got a lot of work to do if we're going to survive on this Eaarth, but most of it needs to be done close to home. Small, not big; dispersed, not centralized.

And so we're left with a big national government and smaller national purposes. Which is the worst place to be, because conservatives are correct about the inherent inefficiency of big government. When Jefferson took over as president, the Treasury Department had grown to eighty employees. "For its critics, it was a monster in the making," the historian Ron Chernow wrote. The new president promised "the employment of the pruning knife." And now? Now government "has never had more layers of leaders or leaders per layer," writes the political scientist Paul

Light. "The towering hierarchy diffuses accountability for what goes right or wrong, weakens command and control, and reduces communication to a childhood game of telephone in which messages are distorted at every stop in the chain of command."[26] It's a Monty Python world. "The fastest-spreading titles in the hierarchy involve chiefs of staff of one kind or another . . . the latest innovation in layering," explains Light. "First created at the Department of Health and Human Services in 1981, the title has been spreading laterally and horizontally ever since. The first deputy chief of staff to a secretary appeared in the hierarchy in 1987, followed by the first chiefs of deputy secretaries, administrators, and assistant secretaries in the early 1990s." In fact, by 2004, there were sixty-four "titles open for occupancy," each created by "new ways of commingling key words such as 'principal,' 'deputy,' 'associate,' 'assistant,' and 'chief.'" Thus the principal assistant deputy undersecretary, or the principal assistant to the associate undersecretary, or the associate principal deputy assistant secretary, or the deputy deputy assistant secretary, or the chief of staff to the associate deputy assistant secretary. Imagine, says Light, that you were a nurse at a VA hospital. You "reported upward through seventeen layers," beginning with a nurse supervisor and finishing with the secretary of veterans affairs. (Nine of those layers were occupied by presidential appointees.) The good news is that all of these people are doing a swell job; in surveys almost all federal appointees said they were "very good" or "above average." Their supervisors apparently agreed; of the seven hundred thousand federal employees who were rated in 2001 using a pass-fail rating system, just 0.06 percent failed. Of the eight hundred thousand who were rated on a five-point system, just 0.55 percent were rated as either "minimally successful," or "unacceptable," while 43 percent were rated "outstanding."[27] Heckuva job, Brownie.

When you had a cause sufficiently grand, you could justify such disconnected inefficiency—"saving the free world" was rationale enough to overlook a lot of Pentagon featherbedding. But now we distrust such grandiosity; our biggest projects seem mainly to benefit a few elites (Halliburton in the Iraq war), and centralization seems as much about plunder as progress. In some sense, we're the owners of more national government than we actually can use—sort of like the 95 percent of SUV owners who never Navigated, Explored, or Yukoned off a paved surface. If you're not defending the free world or paving your continent, you probably don't need the two feet of clearance and a V-8 engine. An Escort might do.

Some places around the world still have huge projects. The Chinese, for instance, often justify an authoritarian and far-reaching central government on the theory that they're a young nation desperately fighting to pull people out of poverty. That government has moved mountains, on the same scale we did when settling the continent; it built the Three Gorges Dam, and now it's diverting the biggest rivers of the wet south to water the arid north. These may not be smart projects, but they are undeniably big. The Chinese feel, as far as I can tell, that they're moving ahead, fast—that they're making over their world almost as completely as we remade ours in the nineteenth and twentieth centuries (and doing it in a few decades). The outcome is likely to be tragic; they're running into the same environmental walls as the rest of the planet, as we've seen. (Think of those glaciers in the Himalayas melting, and with them the Yellow and the Yangtze starting to dry.) Still, for the moment, it feels different from contemporary America or Europe. In the summer of 2008, standing by the fence outside the nearly completed Olympic stadium, I counted thousands of people appearing, hour after hour, just to pose against the backdrop of the Bird's Nest. It's hard to imagine any place in America all *that* thrilled to host the Games, or really concerned whether we were topping the medal count. Our sense of excitement about such things is past; we've gone over the top of the roller coaster. If we're waving our hands in the air, it's less from excitement than from fear.

We still have things we need to do, of course; we still have projects. But in the rich world at least those projects no longer center on expansion and growth. (The poor world will be a focus of chapter 4, and among other things it will be clear that their only hope is if we in the north back off.) From now on, we're about *keeping* what we've got. *Maintenance* is our mantra. When President Obama took over in 2009 and needed to stimulate the economy, he and the Congress focused on *repair.* Not erecting new bridges (we've bridged all there is to bridge; that's why we were reduced, in the Bush years, to building bridges to nowhere), but preventing old ones from crumbling. And that project gets ever more important, and ever more difficult, as the planet turns tougher: when the wind blows harder and lightning strikes more often and more rain falls and the sea rises, repair and maintenance become full-time jobs.

Here's another way of saying it: After a long period of frenetic growth, we're suddenly older. Old, even. And old people worry less about getting more; they care more about hanging on to what they have, or losing it as slowly as possible. That's why old people are supposed to keep their money in bonds, not stocks. Growth doesn't matter. Security and stability count more than dynamism. Assuming you've been generally successful—and we have been, creating more wealth than any society that ever was—your goal becomes to husband that wealth. To prolong a decent old age. As a nation, or as a planet, it's not our chronological age that's in question—we're still a "young country," a young species. But between global warming and the end of oil and the economic backwash from both, it's as if we've come down with a chronic disease that slows us down, stoops us over. Now we have to engage in some triage, decide what from our previous life we most want to keep, and how we plan to do it.

This metaphor occupies my mind because in the last couple of years my mother reached eighty and moved out of the big house where she'd lived for decades. Actually, by suburban standards it was no longer a big house—well below average. But too big for her. The projects that had occupied her life—most of all, raising her children—were more or less complete, and she wasn't planning on having more kids. So she moved into a retirement community, to an apartment off the main wing. The most obvious difference was its size; it could hold about an eighth of the stuff crammed into the old house. But it could hold everything she now needed.

So the first point is simple: the size of your institutions and your government should be determined by the size of your project. The second point is more subtle: *The project we're now undertaking—maintenance, graceful decline, hunkering down, holding on against the storm—requires a different scale. Instead of continents and vast nations, we need to think about states, about towns, about neighborhoods, about blocks.* Big was dynamic; when the project was growth, we could stand the side effects. But now the side effects of that size—climate change, for instance—are sapping us. We need to scale back, to go to ground. We need to take what wealth we have left and figure out how we're going to use it, not to spin the wheel one more time but to slow the wheel down. We need to choose safety instead of risk, and we need to do it quickly, even at the sacrifice of growth. We need, as it were, to trade in the big house for something that suits our circumstances on this new Eaarth. We need to feel our vulnerability. It's not just people in poor nations who are exposed to the elements now, but all of us. We've got to make our societies safer, and that means making them smaller. It means, since we live on a different planet, a different kind of civilization.

I'll spend the rest of this book explaining how we might make that very different world workable—how we might keep the lights on, the larder full, and spirits reasonably high. These are all difficult tasks; the transition from a system that demands growth to one that can live without it will be wrenching. But the most wrenching part will be the simple idea of decline. We don't like aging as individuals, and it frightens us as a society. Ever since Jimmy Carter first hinted at it in the 1970s, we've been desperate to flog our economy back to life. We deregulated, never mind the pollution. We cut taxes, never mind the gross inequality it created. We handed out cheap mortgages, never mind the headache we knew was coming. We have, in short, goosed our economy with one jolt of Viagra after another, anything to avoid facing the fact that our reproductive days were past and hence constant and unrelenting thrust was no longer so necessary. (I suspect global warming is the planetary equivalent of the dread "erection lasting more than four hours" that we're warned about on the TV commercials.)

If I'm a little less paralyzed by this fear, it's for an interesting reason. When I was still a young man, I moved to the Adirondack Mountains of upstate New York, a place notable not just for its beauty, its wildness, and its poverty but also for its history: the population of the Adirondacks, at least where I lived, was far lower than it had been a century before. Its boom had come, and gone, and in fact was all but forgotten. I've written elsewhere about a few days I once spent camped at a place called Griffin, on the upper stretches of the Sacandaga River. If you knew what you were looking for, you could still make out the old cellar holes and find the iron ring bolted into the granite that must once have anchored the mill. In the 1870s this place had a school and a dance hall and long rows of company houses. Every day teams of horses arrived pulling sledges piled high with hemlock bark, the source of the tannin that transformed hundreds of thousands of hides into soft and supple leather. There are photos of the people who lived there once, and they've left behind their stories. (According to one local history, a band once came from Northville, twenty-three miles away, to play at a celebration, and the lumberjacks who were listening to the music "started boxing, which was quite a hobby in those days. When they stopped playing all the men stopped boxing; when they started playing again most of the lumbermen started boxing again. The musicians had a hard time to keep from laughing.")[28]

But then, at the end of the nineteenth century, some scientist discovered a synthetic chemical that could tan hides more cheaply, and within a few years Griffin was abandoned, as was Arietta, and Jerden. And not just abandoned. In the rainy East it didn't take long for rot to topple the old mills and homes, for rust to corrode the old barrel hoops, for birch to sprout in the old cellars and fill them with leaves. Most of the Adirondacks was like that—rocky farms abandoned as soon as the topsoil of the Midwest made them obsolete. The Adirondack stream I described in the preface to this book, and in *The End of Nature*, now flows unimpeded thirteen miles to reach the Hudson, but once it passed through twenty-six mills along that stretch. The Hudson itself once had massive log drives, with rafts of trunks on their way to the mills at Glens Falls—I've met old men who worked on them when they were young. Now the only rafts are rubber toys for white-water tourists, and those mills are mostly quiet. *And it's okay.* Those lumberjacks lived a good life, and the forest that replaced them lives a good life, too. Then millhands; now moose. Though I taught Sunday school in our tiny Methodist church, it always seemed to me a Hindu place—not the linear Western arrow, but the turning Eastern wheel. Its sweetness inoculated me a little against that deep fear of decline.

Still, in less philosophical moments, there is much to worry us. Our ancestors, and we ourselves in the decades just past, piled up a great deal of wealth precisely by ignoring the finite nature of the planet. We also, through that willful ignorance, simultaneously wrecked the prospects for future growth. So we are heir both to the wealth and to the increasingly degraded planet it came from. We have to make that wealth last us. We had better not squander what inheritance we still have, and we had better figure out how to share some of it with the people already suffering from the environmental woes our profligacy caused.

That means reshaping our society. As I've said, growth and expansion seemed to require a kind of centralization—a concentration of resources. We had to be able to pool the continent's capital to build the railroads and highways. And we had to diminish the barriers of sectionalism and local self-interest. Tennesseans, for instance, had to help pay for the railroad across Montana, secure in the knowledge that it would eventually enrich them, too. The central government needed to be able to override the local. Imagine building the Interstate Highway System without the power of eminent domain—without the power to tell a million farmers "tough." My great-aunt in West Virginia was able to make the planners build a bridge over the new road for her cows, but she wasn't able to stop the road. In fact, she didn't want to stop it; she understood it to be Progress.

Maintenance, on the other hand, requires dispersing resources. It's too risky to bet on a few things, because if we lose we can't recover. And on the Eaarth we've created, the odds of losing go steadily up. New Orleans bet on a couple of seawalls. Instead, we need to take a million small wagers, spread the risk as widely as possible. We shall see, in chapter 4, how we've let our energy systems and our food systems grow "too big to fail," just as we did our banks. And the answer is the same—not bigger banks, but smaller banks, small enough that their failure (if it happens) can be absorbed. Food that comes from closer to home, not through an endless and vulnerable chain. Energy from your roof or your ridgeline—energy that doesn't yield quite the power of a barrel of oil, but that doesn't require an army to keep it flowing. *Our* Projects, if we are wise, will be myriad and quiet, not a grand few visible to the whole world.

Should this happen—and on an uphill planet I think it's almost inevitable—the power that's concentrated in Washington will begin to tilt back toward lower levels of government. It will be a more Jeffersonian future than a Hamiltonian one. Which would be ironic, since in Barack Obama we've finally got a president using centralized power to good ends, but even he can't stand athwart the tides of history, especially once the sea has risen a few feet. The most radical words in that most radical document, the Declaration of Independence, are right at the beginning: "When in the course of human events." Times change, and when they do we must respond.

When my wife and daughter and I left the Adirondacks, we moved about fifty miles to the other side of Lake Champlain and the quite different state of Vermont. Different in topography—Addison County, where we live, is like a little chunk of Ohio that somehow ended up in New England, a broad flat valley filled with dairy farms. Different in climate—not quite as harsh. I'll use several illustrations from present-day Vermont in the latter parts of this book. That's partly because I live there; these are the stories of my daily life. But it's also because Vermont is a particularly good example. Not because it's typical—it isn't, which is why I'll also include lots of example from cities and suburbs the world around. But precisely because it's *different,* most of all in its history. In fact, little Vermont may be one of the best places in the world to think about the scale of the future, simply because it's had a very odd past.

The state was created in July 1777, when twenty-eight towns declared their independence from New York. It was, at root, a real estate dispute. New Hampshire and New York had disputed the ownership of the land, and in 1764 the King's Privy Council ruled for New York, setting its eastern boundary on the Connecticut River. But that meant that lots of people holding titles to their farms granted by New Hampshire were suddenly out of luck. They began to agitate. After Ethan Allen and the Green Mountain Boys won one of the first key victories of the Revolution, seizing the cannons at Fort Ticonderoga and humping them overland to Boston to pester the British, Vermonters expected to be welcomed into the new union. But New York wouldn't relinquish its claims, and Congress wouldn't move against New York, and so for fourteen long years Vermont was its own republic, always under threat and always threatening. Its leaders, Ethan Allen in particular, were charismatic, violent, and often drunk—but also shrewd and creative. They treated with the British, mostly to worry the Americans: as Ethan's brother Ira once said, the British represented a north pole, and the Americans a south pole, and "should a thunder-gust come from the south, they would shut the door opposite that point and open the door facing on the north."[29]

One could tell many stories from that time—the interloper who was tried by a vigilance committee and then hung in an armchair from a tavern sign for two hours at a height of twenty-five feet, facing the border of the Empire State, "to the no small merriment of a large crowd of onlookers"[30]—but in the end the intractable Vermonters wore down New York, which relinquished its claims, and Vermont joined the Union as the fourteenth state. For a decade and a half, however, it had been its own free place, not a former colony with aristocratic oligarchs, not a part of the American confederacy, but, as one Vermont historian puts it, "the only true American republic, for it alone had truly created itself."[31] It suggests, in other words, the path the American continent might have taken if it hadn't become preoccupied with all those National Projects.

Vermont played a role in the nation's later, larger history—John Deere forged the plow that broke the plains in a blacksmith shop on the edge of the Middlebury campus—but as a result of its peculiar history, writes the historian Peter Onuf, "nowhere in America did local communities become so thoroughly accustomed to such a high degree of political self-determination."[32] It's a cussedness that has persisted from the start: Vermont was the first state to outlaw slavery, in its 1777 constitution, and the first to let every man vote. Overwhelmingly white, it can nonetheless claim the first black college graduate in the country (Alexander Twilight, Middlebury, 1823) and the first black elected official; more to the point, it defied the federal Fugitive Slave Law by freeing any slave that crossed its borders. More recently, it passed the first civil union law in the nation and became the first state to allow gay marriage without the prompting of a court, but it's not libertarian. It also boasts the toughest restrictions on billboards and land development (and our capital, Montpelier, has fought for decades to retain its status as the only state capital without a McDonald's).

Vermont was always patriotic—it lost the highest per capita

share of its population in both the Civil War and the war in Iraq. But its patriotism was never as enmeshed in the great National Projects as most of the country. In fact, the proudest moment in its history may have come in the 1930s, amid the wreckage of the Depression. In an attempt to grab a share of money from the National Recovery Administration (what today we would call "stimulus money"), local boosters conceived of a plan to build a highway along the crest of the Green Mountains, just like the Blue Ridge Highway in the southern Appalachians. Not only would the new road mean thousands of new jobs; it would also—or so promised its supporters—expose narrow and conservative Vermonters to grand vistas on a daily basis, thus curing them of what the secretary of the state chamber of commerce called their "valley-mindedness." The backers pointed out that Washington was ready to kick in $18 million if the state would just pony up $500,000, and the state legislature narrowly agreed—but with one proviso: the idea would have to pass muster with citizens when the state's 248 towns held their annual town meetings in March. Final vote: 31,101 yes; 43,176 no. I went for a ski this afternoon on the Forest Service track where the highway would have run. I crossed bobcat tracks and saw a moose.

Perhaps because of Vermont's odd history, there have always been some residents who wanted it to break away again and set off on its own. (Others have had the same idea; in 1854, when Vermont officially endorsed the abolition movement, the Georgia legislature passed a resolution recommending that the Green Mountain State be "made into an island and towed out to sea.") We've got a secession movement at the moment that, depending on how you manipulate the polling question, draws support from a fifth of the state's voters. (In fact, across the United States 18.2 percent of people say they would "support a secessionist effort in [their] state.")[33] Some of Vermont's secessionist leaders enjoy dressing up as Ethan Allen and are forever holding conferences with southern diehards who can't wait to return to the old Confederacy, or with Sarah Palin's friends in the Alaskan Independence Party, or with the supporters of Texas soreheads like Governor Rick Perry, who proclaimed early in the Obama years that "if Washington continues to thumb their nose at the American people, you know, who knows what might come out of that."[34]

Too many of Vermont's official secessionists insist that 9/11 was a government job, and revile Abraham Lincoln precisely because he successfully defended the Union (a slur on the many thousands of local tombstones that carry names like Antietam or Bull Run). They undermine, with their conspiracy theories, the idea of independence. But many of my neighbors are de facto parts of the quieter movement for what might be called "functional independence," the people working hard to figure out how Vermont might one day grow more of its own food and provide more of its own energy, deliver most of the goods and services necessary for a dignified life. They campaign against federal subsidies for big agriculture and work to build local food networks; against huge power plants and for laws to make windmills easier to build. There are people like this emerging in every state, and in every county, and in suburbs and cities as well as the countryside. You'll meet some of them in chapter 4. They're enterprising and canny. Most of all, they're *connected*. They sustain *communities*.

Community may suffer from overuse more sorely than any word in the dictionary. Politicians left and right sprinkle it through their remarks the way a bad Chinese restaurant uses MSG, to mask the lack of wholesome ingredients. But we need to rescue t; we need to make sure that *community* will become, on this tougher planet, one of the most prosaic terms in the lexicon, like *hoe* or *bicycle* or *computer.* Access to endless amounts of cheap energy made us rich, and wrecked our climate, and it *also made us the first people on earth who had no practical need of our neighbors.* In the halcyon days of the final economic booms, everyone on your cul de sac could have died overnight from some mysterious plague, and while you might have been sad, you wouldn't have been inconvenienced. Our economy, unlike any that came before it, is designed to work without the input of your neighbors. Borne on cheap oil, our food arrives as if by magic from a great distance (typically, two thousand miles). If you have a credit card and an Internet connection, you can order most of what you need and have it left anonymously at your door. We've evolved a neighborless lifestyle; on average an American eats half as many meals with family and friends as she did fifty years ago. On average, we have half as many close friends.

I've written extensively, in a book called *Deep Economy,* about the psychological implications of our hyperindividualism. In short, we're less happy than we used to be, and no wonder—we are, after all, highly evolved social animals. There aren't enough iPods on earth to compensate for those missing friendships. But in this book I'm determined to be relentlessly practical—to talk about surviving, not thriving. And so it heartens me that around the world people are starting to purposefully rebuild communities as functioning economic entities, in the hope that they'll be able to buffer some of the effects of peak oil and climate change. The Transition Town movement began in England and has spread to North America and Asia; in one city after another, people are building barter networks, expanding community gardens. And they've paid equal, or even greater, attention to suburbia; in the developed world, after all, that's where most people live. Though our sprawl is designed for the car, the sunk costs of those tens of millions of houses mean they're not going to disappear just because the price of gas rises. They'll have to change instead. "Suburbia, not as a model for material consumption, but as a legal and social lattice of decentralized and more uniformly distributed production land ownership, has the potential to serve as the foundation for just such a pioneering adaptation," writes Jeff Vail, a widely read economic theorist who envisions "a Resilient Suburbia."[35]

In fact, quite sober economists have begun to insist that even in our seemingly globalized world, our economies are actually far more local than we realize. Despite the "pervasive image of a single U.S. economy," the economists William Barnes and Larry Ledeber write, "local economies—primarily metropolitan-centered and strongly linked—are the real economies in the United States." They build, with rich statistical backing, on the original insights of thinkers like Jane Jacobs, who always insisted that the city was the fundamental building block of our economic life. These "Local Economic Regions" comprise the web of transportation and communication links, the chain of educational institutions in a region, and the web of emotional ties. (My Vermont neighbors may not care much how many gold medals the United States captured at the Olympics, but they are deeply involved with how many runs the Red Sox scored last night.) Those local economies were originally shaped by geography—a port, a river, a low place in the mountains where you could build a canal. For a while those assets seemed less important; with endless cheap energy, you could always put something on a truck or a plane. But the cities built on those early patterns persisted; they were a sunk cost, too. No one was going to move Buffalo, with its museums and universities and square miles of housing stock, just because the highway had bypassed the Erie Canal. (And now some of those original assets may be returning to prominence. The Erie Canal, for instance, has seen a marked upswing in business as the price of oil rises, because a gallon of diesel pulls a ton of cargo 59 miles by truck, but 514 miles in a barge.)[36] Shanghai is 7,371 miles from New York. It's true that Chinese workers cost you a dollar an hour, but at some point the math shifts.

Even David Ricardo, the nineteenth-century economist who helped kick off globalization with his theory of comparative advantage, never quite imagined the Flat Earth we've lately celebrated. It was true, he said, that since Britain could make cloth more cheaply, and Portugal wine, each country should specialize. He believed, however, that capital would stay at home, due to "the natural disinclination which every man has to quit the country of his birth and connexions and entrust himself, with all his habits fixed, to a strange government and new laws. These feelings, which I should be sorry to see weakened, induce most men of property to be satisfied with a low rate of profit in their own country, rather than seek a more advantageous employment for their wealth in foreign nations."[37]

David Ricardo, meet Woody Tasch. A New Mexico–based venture capitalist and the founder of the Slow Money movement, Tasch focuses on finding funds to help local businesses grow a little larger. Not the kind of money that's looking for a 20 percent annual gain; when that happens, everything but return gets pushed aside. What Tasch has in mind is a consistent, sound, 3 or 4 percent return, which at the same time benefits the community where both the investor and the business live. "These kinds of local businesses are by definition going to be lower risk, because they're embedded in their communities, they're cooperating with each other," he says.[38] They can use those networks to grow, but only up to a certain point—and you only *want* to grow to a point. Ben and Jerry's was great when it was a Burlington ice cream shop, and pretty neat when it was a regional brand—but now it's owned by Unilever. What if your newspaper wasn't owned by some corporate overlord looking for a 20 percent return? What if a small annual profit was enough? Maybe it would still be covering the city council and sending a reporter on the road with the baseball team.

But in our world, it's actually harder than you'd think to stay small. To understand why, visit the Farmers Diner, one of my favorite restaurants but also a place that illustrates just how hard it can be to find the sweet spot. How local is the Farmers Diner? The first thing you see when you walk in the door of their outlet in the Vermont town of Quechee is a jukebox, glinting like any diner jukebox. Some Willie Nelson, some John Cougar Mellencamp. But half the albums are by Vermonters. Phish, sure. But it's Grace Potter and the Nocturnals who get the most play. And they're just the start. You'll find the Starline Rhythm Boys (singing "The Tavern Parking Lot") and Banjo Dan and the Mid-Nite Plowboys ("The Cider Song"). And Patti Casey, of course. Never heard of Patti Casey? Your loss, but that's the point. In an economy where music comes from L.A. or Nashville, she's from here.

The menu, at first glance, looks like any diner menu. Hash and eggs. Liver and onions. Bacon cheeseburger. Pancakes. At diner prices: $5 for a grilled cheese, home fries for $1.75. But look a little closer: almost every item comes with a modest biography. The blue cheese comes from Jasper Hill Farm in Greensboro. The yogurt is from Butterworks Farm up in Westfield, which also supplies wheat flour for the pancakes. In an economy where diner food rolls up on an eighteen-wheeler from the factory farms of the South and Midwest, your Farmers Diner patty melt is like the music on the jukebox: it comes from here.

And it comes with an attitude. One page of the menu is given over to the Kentucky farmer and writer Wendell Berry's magnificent poem "Manifesto: The Mad Farmer Liberation Front": "So, friends, every day do something / that won't compute . . ." Another is taken up by Thomas Jefferson's 1803 letter calling for a conversion of the nation's "charitable" institutions into "schools of agriculture" so our citizens may "increase the productions of the nation instead of consuming them." This may be the only diner in the world that comes with a mission statement: "to increase the economic vitality of local agrarian communities." The bumper sticker above the counter says it even more plainly: "Think Globally—Act Neighborly."

But it also comes with a problem. In the words of the owner, Tod Murphy, "How do you create a company that will take food off the farmer's hands in the easiest way for him, and set it in front of the customers in the easiest way for them, and do it at a price point everyone can live with?" Tailing him for a day as he made the rounds of his suppliers shows both the promise and the difficulty of the idea. You could start the morning in Strafford, say, at Rock Bottom Farm, where Earl Ransom's cows were producing organic milk and cream on the land where he was born. "I had to educate people that cream isn't necessarily white," Murphy recalled. "When the cows went out to pasture in the spring, the half-and-half changed color noticeably, and the waitresses were afraid people would freak."

It doesn't always go so easily, though. Consider, for instance, the pig. When the first Farmers Diner opened in Barre, it needed bacon—you can't have a diner without bacon. The problem was that no one was producing pork commercially in Vermont. Fifty years ago, sure, every farm had a few hogs growing fat on leftover milk from the dairy herd. But as agriculture became a commodity business—as dairy producers concentrated on cows, and pork producers on pigs—that changed. Vermont dairies became fewer in number and much, much bigger; in other parts of the nation the same thing happened with hogs. According to Brian Halweil in his book *Eat Here*, there's a hog farm in Utah with 1.5 million pigs.[39] That's absurd—the pigs produce more solid waste each day than the entire city of Los Angeles. But it's also cheap—so cheap that it sets the psychological price for a pound of bacon pretty low. So when Murphy wanted to buy pigs for his bacon and sausage, he approached a few farmers to see whether they were interested. One was Maple Wind Farm, a breeder in Huntington raising fifty hogs a year, mostly to sell at farmers' markets. They're fed on grass and organic grains—the pork tastes absolutely incredible—and they fetch good money. "We get $7.50 a pound for bacon at the farmers' market, and $8.50 a pound for pork chops," says Beth Whiting, who runs the farm with her husband, Bruce Hennessey. So when Murphy asked them if they could raise him some pigs at eighty-nine cents a pound, "we had to bury our laughter."

And yet eighty-nine cents a pound is more than the upscale national pork producer Niman Ranch pays its contract pig farmers. In essence, it's a Goldilocks problem: somehow Murphy has to find just the right size. What his operation really requires is not huge commodity producers or small, incredibly wonderful gourmet farms. "What I need are 1950s-size farms," he says. Not a million hogs, but not fifty, either—maybe three or four hundred. Not organic operations necessarily, just family farms. Precisely, in other words, the kinds of farms that have almost all gone out of business in recent decades.

Murphy can still find vegetable growers to fit his scale, for example, someone to plant the five acres of cucumbers he needs for his pickles. But to help rebuild the supply of meat and chicken farmers, he's launching a nonprofit foundation. Named for a character in one of Wendell Berry's novels, the Jack Beecham Foundation will help growers with business plans and marketing strategies. Woody Tasch has been helping.

All this to make a smoked-turkey club. Or, to read from today's specials menu, some poached Vermont eggs with Cabot cheddar cream sauce. Or some maple butternut squash. Or some Cortland apple cobbler topped with local granola, and a scoop of that Strafford ice cream. With some Grace Potter wailing from the jukebox. For change back from a ten-dollar bill, it doesn't get much sweeter than this. It should work. It should spread. If the eaarth is going to support restaurants, they'll need to look like the Farmers Diner.

Across the country communities have begun to transform themselves. They encounter the same kinds of problems that trip up Murphy, but they find solutions, too. Often a farmers' market is the catalyst—not just because people find that they like local produce, but because they actually *meet each other* again. This is not sentiment talking; this is data. A team of sociologists recently followed shoppers around supermarkets and then farmers' markets. You know the drill at the Stop'n'Shop: you come in the automatic door, fall into a light fluorescent trance, visit the stations of the cross around the perimeter of the store, exit after a discussion of credit or debit, paper or plastic. But that's not what happens at farmers' markets. On average, the sociologists found, people were having ten times as many conversations per visit. They were starting to rebuild the withered network that we call a community. So it shouldn't surprise us that farmers' markets are the fastest-growing part of our food economy; they are simply the way that humans have always shopped, acquiring gossip and good cheer along with calories.

And it can easily go beyond food. Bellingham, Washington, is a good example. The local chapter of the Business Alliance for Living Local Economies has five hundred member merchants, and 60 percent of the city's eighty thousand residents tell pollsters they've changed their buying habits dramatically. Grand Rapids, Michigan, couldn't be any more Rust Belt and any less hippie—its most famous product was Gerald Ford. But in three years 250 businesses have joined the Local First campaign. "Lots of people have lived here all their lives," Guy Bazzani, a local green architect and founder of the network, told the *New York Times*. "When you have a generational community like this, there is a lot of natural social responsibility. You don't have to call it that. But people just love the community and they'd rather buy a product from someone they know."[40] You get all that warm fuzzy stuff—and you also avoid the risk that comes with a Wal-Mart economy. Not just the climate change and the peak oil, but the risk that Wal-Mart is going to gut your downtown and then pull out. It's the largest owner of vacant buildings on the planet.

The most avant-garde adventures in localizing economies may be the proliferating local currency projects—four thousand of them around the world. Most of these are small, the notes accepted by a few progressive-minded businesses in college towns. (If you want a backrub from a vegan, you may be in luck, but try buying a stapler.) Still, some homegrown currencies are much more elaborate and broad-based. In the Swiss WIR system, which dates to 1934, when a buyer makes a purchase the seller receives a credit in his account, which he can spend, and so on. Similar projects are now under way in Belgium, France, and Germany, says Margrit Kennedy, a Berlin-based consultant.[41] Other such schemes are more tactile. In Lewes, England, for instance, the town began issuing pound notes with pictures of native son Thomas Paine. The *Financial Times* reports that the new currency's effect "on individual emotional well-being" has been even more important than its environmental and economic impact. "A relationship develops between resident and trader," one local merchant explained. "People talk to each other about their lives and we've seen the level of trust grow enormously within the community."[42]

My favorite local currency circulates in the Berkshire Mountains of western Massachusetts. Berkshares come in five denominations, from one dollar to fifty dollars, elegantly printed with pictures of local heroes: a Mohican Indian, the artist Norman Rockwell, Herman Melville, W. E. B. DuBois, and so on. You can get the notes at twelve local banks. Cash your paycheck, and it comes in either greenbacks or Berkshares, or some mix of the two. There are $2 million of the notes in circulation, and you can spend them almost anywhere—I shopped with mine at the organic food co-op, but also at the Mr. Dingaling ice cream truck. You know the world could shift when you can hand over a Norman Rockwell for a Fudgsicle and get a couple of Herman Melvilles back in change.

Embracing the local doesn't mean abandoning the connection to something larger. No one is tossing out U.S. currency, and it will be a while before there's a village computer maker or a local locomotive manufacturer. "Forget a Robinson Crusoe economy," advises the local advocate Michael Shuman. "A self-reliant community simply should seek to increase control over its own economy as far as practicable."[43] And past and present, small economies have done just fine: in his classic book *Human Scale,* Kirkpatrick Sale tells the story of Lucca, the Tuscan city-state that was a "fiercely independent republic" for four hundred years, with a "rich and self-sufficient agriculture," banks, churches, artists, musicians. Only its eventual absorption into the kingdom first of France and then of Italy reduced it to a "forgotten backwater, with a shaky industrial base and an agriculture quite dependent on imports."[44]

I've been in these Tuscan towns, the source of so many American fantasies of the good life. (The source of what can only be called "good-life porn.") I remember sitting in the pew of the twelfth-century Abbazia Di Sant'Antimo, not far from Siena in the hill town of Montalcino, one hot summer afternoon listening to the monks chant Nones in sonorous harmony. I kept looking past the altar to two windows behind. They framed prime views of the steeply raked farm fields in back of the sanctuary: one showed rows of dusty-leaved olive trees climbing a hill, the other rank upon rank of grapevines in their neat rows. With the crucifix in the middle they formed a kind of triptych, and it was easy to imagine not only the passion but also one's cup running over with Chianti, one's head anointed with gleaming oil. And easy enough to figure out why this Tuscan landscape is so appealing to so many. Its charm lies in its *comprehensibility*; its scale makes intuitive, visceral sense. If you climb one of the bell towers in the hill towns of Tuscany, you look out on a compassable world; you can see where the food that you eat comes from, trace the course of the rivers. It seems sufficient unto itself, as indeed it largely was once upon a time.

And in the ancient churches it's easy to construct a vision of the medieval man or woman who once sat in the same hard pew—a person who understood, as we never can, his or her place in the universe. It was bounded by the distance one could travel physically; save for the Crusade years, it was probably easy to live a life without ever leaving the district. (Florentines speak of living an entire life in view of the Duomo.) And it was bounded just as powerfully by the shared and deep belief in the theology of the church. *You knew your place.*

This phrase, of course, has several meanings. You would have been deeply rooted in that world. (It's hard to imagine anyone having the identity crisis that's routine in our world.) But you would have been considerably more rooted than we're comfortable with. You knew your place in the sense that you were born into it, and there was little hope of leaving if it didn't suit. Peasants were peasants and lords were lords, and never the two met. Inequality was baptized, questioning unlikely. The old medieval world made sense, but it was often an oppressive sense—hence the five-hundred-year project to liberate ourselves in every possible way.

And though Tuscany still looks comprehensible—making it the backdrop for profitable tourism and powerful travel fantasy—it's now partly sham. The farms remain, largely supported by farm subsidies from the European Union or the wine-buying habits of affluent foreigners. The villages are mostly emptied out, with only the old remaining; on weekends traffic swells as Florentines and Romans head to the country house. Even the churches are largely relics—stop in for afternoon mass, and you're likely to find three or four old women listening to an African priest limp along in halting Italian—there aren't near the vocations necessary to fill these pulpits. Even the chanting monks at the Sant'Antimo abbey are imports—a French brotherhood that took over the church a decade ago. It's not quite real.

But it's coming back. Tuscany is one of the birthplaces of the Slow Food movement, which is slowly reclaiming those farms, making them appealing again to young people. In fact, small places are in certain ways now more real than the fantasy lands that dominate Washington or Wall Street, even when it comes to huge global projects. At least until the Obama administration, for instance, American states and cities had done far more than the federal government to fight climate change. They made real change, change that mattered. Some of the shifts were small: across the United States, municipalities installed light-emitting diodes in traffic signals to save energy—Philadelphia saves $800,000 a year from reduced electric use. Sometimes individual states are large enough by themselves to force big shifts in policy: the automakers in Michigan were able to stave off fuel efficiency changes in Washington for decades, but California finally forced them to start producing higher-mileage cars and trucks. (If you make a car you can't sell in California, your car company might . . . go broke.) And often states band together: the first effective caps on power plant emissions came from eleven New England and mid-Atlantic states that came together in something called the Regional Greenhouse Gas Initiative, or RGGI.[45]

I'm not suggesting that every state secede from the union, and every town secede from every state, and every neighborhood set off on its own. There's a kind of stability that comes from spreading risk across a continent: New Orleans couldn't have repaired itself. When my town flooded, federal money helped. And giving up on the national government would mean giving up on much that's good. Our National Projects weren't only about paving highways. They were also about guaranteeing civil rights and setting aside wilderness areas, protecting free speech and endangered species. Such advances would fare less well, at least in places, if we broke the country down into tiny slivers. One imagines the Alaska Independence Party, for example, would drill for oil in every square inch of the tundra, caribou be damned. So all I'm suggesting is that, on a hot and difficult planet, decision making will need to start sliding toward more local levels. The key projects aren't national anymore.

And if we're going to keep the roads intact and silos full, we won't be able to keep underwriting some of the distinctly national indulgences we've long supported. The U.S. military, for instance, costs more than the armies of the next forty-five nations combined; the Pentagon accounts for 48 percent of the world's total military spending.[46] You don't need to be a pacifist to wonder if we can still afford that kind of money; instead, you might be a state legislator or a city councillor. In Vermont, by one estimate, we ponied up $150.6 million in 2009 for "oil-related military efforts." If we'd spent that money on, say, renewable electricity, 225,000 Vermont homes would have gone green. Since we have only 240,000 households, that's pretty good.[47] Consider the essayist Miryam Ehrlich Williamson, who lives in a Massachusetts town of 750 people. According to her figures, she and her neighbors paid $1.8 million in 2007 for the war in Iraq, an amount almost exactly equivalent to the municipal budget. That same $1.8 million could have instead bought the services of thirty-six policemen for a year,[48] which is doubtless far more than they need—but certainly enough cops to pretty much *guarantee* their security, in a way the U.S. military really doesn't.

In fact, security is one of those ideas whose meaning begins to shift in a new world. If we figure out how to provide more of our food and energy close to home, the need for the kind of standing army that the founders deplored will begin to decrease—no more need to guard the five-thousand-mile-long straw through which we suck hydrocarbons from the Persian Gulf. If western Massachusetts was generating most of its own energy, western Massachusetts would have less need for that kind of military. But it might need other kinds of protection. If you think about the cramped future long enough, for instance, you can end up convinced you'll be standing guard over your vegetable path with your shotgun, warding off the marauding gang that's after your carrots. (Police have already reported upswings in the theft of solar panels,[49] and in game poaching, with "gangs from town rampaging through the countryside with guns, crossbows, or snares.")[50] The marines aren't going to be much help there—they're not geared for Mad Max—but your neighbors might be. Imagining local life in a difficult world means imagining taking more responsibility not only for your food but for your defense. (Consider Switzerland, for example, where every adult male is a soldier.) *Militia* is an ugly word to many of us, but it's worth remembering, at least for those of us with tricorne hats in the closet, that a local militia fought the fight on Lexington Green.

In the new world we've created, the one with hotter temperatures and more drought and less oil, big is vulnerable. We are going to need to split up, at least a little, if we're going to avoid being subdued by the forces we've unleashed. Scale matters, and at the moment ours is out of whack with our needs. Mammals get smaller in the heat, and so should governments. Our key projects are local now; that needs to be our focus.

It shouldn't be completely scary; we almost intuitively realize that for many things small is often better. (Think about whether you want your child going to a big school or a small one.) When asked where they would most like to live, in survey after survey, most Americans choose "small towns."[51] Carving workable "small towns" out of cities and suburbs—giving people everywhere the control and cooperation they'll need to ride out the storm—is key work for the time to come. It's not impossible; across South America even very poor neighborhoods in huge sprawling cities have taken control of a share of the municipal budget in recent years. New technology, as we shall see in chapter 4, can help. But in the end, the transition will need to be mostly mental: we need to get past our current ideological rigidity and think more broadly. It's not at all clear whether a farmers' market, or a local neighborhood crime watch, or a community-owned windmill is a liberal or conservative project. It's some of both. Mostly it's some of neither—our politics, like our highways, were built for an era of endless growth. Karl Marx as much as Adam Smith thought we'd end up in a material paradise; Richard Nixon and Nikita Khrushchev sparred over whose system would produce better kitchen appliances. In the age now dawning, our hopes will shift and our ideologies will shift with them.

There's every reason to hope that we will be able to find real community, even in the face of disruptive change. In one of the most lyrical volumes of history in many years, Rebecca Solnit chronicled the kind of arrangements that humans work out for themselves—often without any government help—in the face of disasters ranging from the San Francisco earthquake to New York after 9/11. Far more people take care of each other than take advantage of each other, she found. The official fears of looting and chaos are almost always dramatically overblown, and instead people figure out how to feed and clothe and shelter victims. "Paradises built in hell," she called them—temporary utopias, and maybe not so temporary. The world we're now stumbling into may well resemble one long not-so-natural disaster, with troubles piling on top of each other and the easy consumer answers beginning to break down. Such crises, she says, "reveal mutual aid as a default operating principle."[52] It both forces and allows us to be more than consumers—neighbors, citizens.

Still, no use underestimating the depth of change we'll need to deal with, especially since there's no end point in sight. As we lose the climatic stability that's marked all of human civilization, it's not as if we're going to land on some other firm plateau. The changes to our lives will be ongoing and large and will require uncommon nimbleness, physically and psychologically. Our focus will have to shift. As a culture and an economy, we've had the margin to afford a lot of abstractions. Abstractions in the supermarket aisle: Lunchables, and Cheetos, and the four thousand other incarnations of high fructose corn syrup. Abstractions in our relations with the rest of the planet: "the free world." Abstraction will grow harder; increasingly, we're going to have to focus on essentials: on actual food and on energy that comes from the wind and sun in our neck of the woods, not from that abstraction called "the Middle East."

We're going to have focus on work, not "work." It's true that we're spending more time at the office than ever—leisure hours have fallen about a third since 1950.[53] (By contrast, hunter-gatherers "work" about two hours a day, and even at peak harvest times Amish farmers average just eight hours a day.)[54] But a lot of what we're doing isn't exactly real. Not long ago, the *New York Times Magazine* organized a forum that unintentionally but brilliantly defined the age now ending. "Let me give you the scenario," the writer said to several Madison Avenue geniuses. "I'm the somewhat desperate CEO of a company called Jack's Overalls. We manufacture functional clothes, and in the era of corporate farming, our market is fading. My younger vice presidents are telling me that we need to try new media, so I'm turning to you."

> LARS BASTHOLM [chief creative officer at AKQA, where he works on the Xbox account]: Overalls are a staple of Americana, a cultural icon. The question is, how can you make overalls relevant to people today. . . .
>
> BENJAMIN PALMER [chair of the Barbarians Group, which created the "Subservient Chicken" campaign for Burger King]: So you have to create a new market. Farming may be going away, but what's on the rise? Right now your overalls are made with special pockets and holders for farming tools. Maybe we retool them for urban farmers, as it were, with their special-

ized gear. You have special pockets for your iPhone, and your Blackberry, and a pocket for your headphones. . . .

BASTHOLM: Let's really take the brand into the 21st century, shall we? Why don't we put a ShotCode on the front of every single pair of overalls. A ShotCode is like a bar code. You scan it with the camera in your cellphone. And then something comes out the other end . . . it could be a song, it could be a picture, it could be a link to a Web site.

REPORTER: People would come up and shoot me with a cellphone?

BASTHOLM: Yes, with a phone camera. . . . The ShotCode might take people online to a new Web site you've selected. . . .

ROBERT RASMUSSEN [he's worked for Nike, Sega, and JetBlue]: Maybe that's something you do in partnership with Facebook or MySpace.

PALMER: . . . Facebook users can buy the Facebook edition of these overalls. They come precoded with your Facebook page embedded. . . .

RASMUSSEN: I would recommend a Web presence built around a utility that engages consumers and allows them to take your brand and own it. . . . Maybe you create a widget that lets you drag your overalls and drop them onto an existing image . . . so you can say, "Boom, that's how it would look if I wore a pair of cord overalls with a blue jean jacket."

BASTHOLM: Yeah, create a little viral engine called You Need Overalls, where you can take current events and just drag a pair of overalls onto whoever's in the news

RASMUSSEN: . . . There's Barack Obama, wearing his jacket and a pair of overalls, giving a speech. . . . Maybe you have samples of these user-generated images playing on television screens in cabs and on digital billboards. . . . You can post the images on your Web site. Create a gallery that shows how overalls can mesh with many styles, from metro to hip-hop to blue collar. People can comment and vote on their favorites.

BASTHOLM: Maybe we create a site called Overall This. Send in a picture of somebody and get them back in overalls.[55]

Or maybe it's you in the overalls. And maybe they've got some dirt on them. Adam Gopnik, reporting recently from the front lines of evolutionary theory, found that scientists now believe that extravagant plumage and giant antlers are the products of long eons of calm like the one our civilization has grown up in, when there's plenty of margin to let a thousand peacocks bloom. "The variations of abundance die at the moment of crisis, and the old stable dull solutions come to life again," he observes.[56] Evolutionary pressure produces peahens, not peacocks; overalls, not Shotcodes. And so we turn to the essentials of our future. In order: food, energy, and—yes—the Internet.

· 4 ·

LIGHTLY, CAREFULLY, GRACEFULLY

Kip Cullers personifies the case for industrial farming. A tall, thin Missouri farmer (who wears, as it happens, Big Smith brand overalls), he holds the world record for growing the most soybeans per acre—154 bushels, in 2007. The average soybean farmer would need the bed of a single pickup to haul away an acre's worth of harvest; Kip Cullers would need four. "They've started calling me the Bon Jovi of farming," he says a little bashfully.[1] Cullers stars in ads for Pioneer Hi-Bred seeds and DuPont herbicides, and opens his farm for a giant industry-sponsored field day each fall so other farmers can see his secrets, most of which involve incredible amounts of chemistry and petroleum. "What Kip does, to get early season weed pressure out of the way, is spray an herbicide before he plants," one BASF sales rep explained. According to the company, Cullers makes use of the entire "BASF portfolio of crop protection innovations," ranging from Respect and Warrior insecticide to Status, Prowl, and Extreme herbicides. (The company even let him spray its new Kixor brand herbicide before it was registered by the EPA.) His field day in 2007 featured a small tub with four stalks of Optimum GAT corn, an as yet unreleased Pioneer variety that has been genetically modified so that farmers can spray even more herbicide without damaging the crop.

Cullers, whose farm fields stretch across seven counties, drives eleven thousand miles in his pickup during spring planting just checking on his fields. He has fifteen tractors, the biggest of which steers itself with GPS satellite data and retails for $185,000. He irrigates his soybeans early in the season—two or three tenths of an inch of water every day, starting on the Fourth of July.[2]

To his corporate backers, Kip Cullers and all that he represents is the key to a peaceful planet. "We have to feed the world," William Niebur, Pioneer's vice president of research and development, told Bill Donahue of *Wired* magazine, who wrote an admiring profile of Cullers in that journal usually devoted to hard drives and virtual reality. "And we can, by increasing productivity per acre. And if we bring people food there will be political stability, which will lead to economic growth."[3] History backs him up: technical changes in agriculture have led to enormous gains in yield, in productivity, and—bottom of all bottom lines—in calories per person. Malthus warned that population growth would outstrip food production, but for a very long time Malthus was wrong. At the beginning of the eighteenth century, the typical French diet contained about 1,850 calories a day. "At that level of nutrition, even the strongest males ha[d] limited vigor for work," one researcher explains, because the sheer demands of metabolism used up most of their energy. By 2007, the average American male had 2,600 calories worth of energy left over each day after his metabolic needs had been met (and since he was likely to be sitting at a desk and staring at a screen, that explains the tummy).[4]

In the years after World War II, even as human populations

skyrocketed, global grain harvests double-skyrocketed. Thanks mostly to Green Revolution technologies of the type Cullers helps to develop, and thanks to the "get big or get out" policies of one government after another, the amount of grain grew from 285 kilograms per person in 1961 to a peak of 376 kilograms in 1986. But since then, grain yields have begun to stagnate, while population has kept growing; by now, the average human is back to 350 kilograms per year.[5] Read those numbers again. For the last quarter century, despite the rapid spread of massive-scale corporate agribusiness farming, despite the help of Warrior and Extreme and Prowl and Respect and Kixor, despite the advent of genetically engineered crops, despite the $185,000 tractor, the amount of food per person has been dropping. The amount of stockpiled grain on the planet, the stuff sitting in silos and warehouses to help us through rough patches, has fallen from 130 days' worth of eating in 1986 to about 40 days' worth in 2008.[6] Forty days sounds almost biblical. So, too, do the food riots in thirty-seven countries, and the rapid rise in malnutrition, which added 75 million people to the rolls of the malnourished in 2007. That is, *the number of people with too little to eat is now rising instead of falling, and rising fast.* In 2000, at their Millennium Summit, the world's leaders pledged to cut in half the number of hungry people in the world by 2015. Just a few years later, the head of the UN's Food and Agriculture Organization revised the timetable, saying that at the current pace even that modest goal would take until 2150.[7]

That's bad news. And it comes before we feel anything like the full effects of what we've done to our planet. Consider global warming. For years, those determined to be cheerful about climate change (mostly utility executives) insisted that a warmer planet would allow us to grow more food. "Look for Siberian wheat in your favorite ciabatta any day now," one tongue-in-cheek columnist predicted.[8] In fact, though, the science now makes clear that "far from compensating for the other damages associated with climate change, hotter temperatures will seriously diminish the world's ability to feed itself."[9] Some of that damage will come from the direct effect of carbon dioxide—in summer 2009 German agronomists found that wheat grown under the levels of carbon dioxide we expect by midcentury will contain markedly less protein and iron and 14 percent more lead. "In addition, it was worth way less money, with smaller grains that are harder to sell for good prices."[10]

But the simplest reason to fear for our dinner is that it's getting too hot for our crops—which, remember, have evolved over deep time to grow on the old earth. Every climate model we've got shows that "by the end of the century an average July day will almost certainly be hotter than the hottest heat waves we experience now," according to a team of American researchers. And those hottest heat waves decimate our fields. in Italy and France, for example, the 2003 heat wave that killed tens of thousands of people also cut corn yields by a third. "It simply becomes too hot for growing plants," says Rosamond Naylor, the director of the Program for Food Security and the Environment at Stanford. "The heat damages the crop's ability to produce enough yield."[11] A 2009 study found that a million square kilometers of Africa might soon be too hot to grow crops. (That's an area larger than the United States plants in its eight largest field crops combined.) "Maize will basically no longer be possible" to cultivate across wide swaths, the study found; there will simply be too many hot days.[12] Pioneer and Monsanto will doubtless try to produce something nifty—Reactor and SunStar and HeatShield and RayFighter—to cope with the rising temperatures, but that will work only to a certain point. In the tropics, and subtropics, the new studies predicted, there's a 90 percent chance that by

2100 the average temperature in the growing season will be hotter than "any temperatures recorded there to date." Past a certain point, corn won't fertilize, rice won't grow.[13] And people can't work, at least not as hard. A new study from the Australian researcher Tord Kjellstrom found that by 2030 Indian laborers would be 30 percent less productive, simply because of increased heat. As one Bengali farmer put it, "Working under the open sky during summer has become nearly impossible—for farmers and their cattle alike."[14]

These changes are already kicking in. As we've seen, since warm air holds more water vapor than cold, evaporation is increasing, and since what goes up must come down, rainfall is increasing too. "Extreme precipitation events" across the United States have increased 36 percent as we've warmed the climate.[15] Instead of the steady and predictable showers of the past, "rainfall has been coming in fits and starts—dry spells interrupted by deluges," reports the *Newsweek* environmental correspondent Sharon Begley. "That is a prescription for withering crops and then washing them away."[16] Eons-old farming practices suddenly don't work. "We've stopped seasonal planting," explained one small farmer in Uganda. "Now we just try all the time. We used to plant in March and that would be it. Now we plant again and again. We waste a lot of seeds that way, and our time and energy. . . . Sometimes you feel like crying."[17] Droughts have returned to the United States, too, even though people have stopped wasting so much water in their homes. The federal government says that thirty-six states face water shortages in the next five years,[18] which is bad news for farming, since 70 percent of the water we use goes for irrigation, and irrigated fields supply as much as 40 percent of the world's food. By 2007 half of Australia's farmland was in a declared drought, and a farmer was committing suicide every four days.[19] In California, in the spring of 2009, groups of farmworkers, many wearing surgical masks against the blowing dust, marched for four days to demand the federal government somehow supply them with more water—the year's drought had already cost the state 23,700 jobs and $477 million in revenue.[20] Farmers are already letting orchard trees die for lack of water to keep them alive; in the Central Valley unemployment grew 9.4 percent in a year through July 2009. "There's no water, so there's not much work," Kiki Torres told a reporter.[21]

As we melt those glaciers and snowfields in the continental interiors, things will only get worse. In the first major speech he gave after being named secretary of energy, the Nobel laureate Steven Chu told an audience in his native California: "I don't think the American public has gripped in its gut what could happen." If we don't dramatically slow global warming, he said, the rapid melt of the Sierra snowpack means "we're looking at a scenario where there's no more agriculture in California," adding, "I don't actually see how they can keep their cities going," either.[22] That's not the kind of thing cabinet secretaries usually say. In the Tibetan plateau, where the Yangtze, Yellow, Ganges, Brahmaputra, and Mekong rivers flow out of rapidly dwindling glaciers, no one knows quite how the farmers downstream will cope. Nor how farmers anywhere will manage as global warming increases pests: it's not just mosquitoes bearing human diseases and bark beetles killing trees that like our hotter planet, but the European corn borer, and the western corn rootworm, and the corn earworm—scientists at Purdue University recently predicted that all would dramatically expand their range in the years ahead. "The warmer readings could mean the insects will be capable of producing up to three generations of their kind in a single growing season, filling fields with their hungry offspring," explained Noah Diffenbaugh, the researcher who led the study.[23]

Remember, too, that other main feature of our new planet, its rapidly emptying oil fields and gas wells. This decline wreaks its own kind of havoc, since the modern agriculture on which we've staked our future requires massive amounts of energy: to make the fertilizer that nourishes the fields, to run the enormous equipment, to dry the crops before they can be stored, to truck the food around the continents. It takes the equivalent of four hundred gallons of oil annually to feed an American, and that's before packaging, refrigeration, and cooking.[24] In 1940, our food system produced 2.3 calories of food energy for every calorie of fossil fuel it consumed. Now, says Michael Pollan, "it takes ten calories of fossil energy to produce a single calorie of modern supermarket food. Put another way, when we eat from the industrial food system, we are eating oil and spewing greenhouse gases."[25] When we try to address one problem, the other gets worse—one reason starvation is on the rise is that the United States now uses a huge chunk of its topsoil to grow gasoline, not food. And not just here. In Ethiopia, where the fast rise in oil prices caused economic devastation, the government tried to compensate by allocating 988,000 acres of land for growing castor beans for biofuels. "Castor plots have so rapidly expanded that they are even depriving us of space for graveyards," one local official said. Ashenafi Chote, a farmer whose small plot used to produce 220 pounds of corn a year, signed on for the beans instead; when the multinational that was supposed to buy them couldn't get the necessary loans, he and his family began to starve. "I made a mistake," he said.[26]

It's hard not to think we all made a mistake. Because yields were rising, we never took seriously all the warnings. In much the same way that rising house prices lured people into ever deeper debt, the Green Revolution lured us into a kind of ecological debt we're only starting to comprehend. As Graham Harvey, the English food researcher (and the farming consultant to the radio drama *The Archers*), points out, "Our food supply is now more dependent on globally traded grains than at any time in our history. This makes it inherently unstable and vulnerable to the kind of catastrophic meltdown that threatened the banking industry." Tommy Thompson, who ran the Department of Health and Human Services for George W. Bush, remarked as he was leaving the job, "I, for the life of me, can't understand why terrorists haven't attacked our food supply, because it is so easy to do so."[27]

Of course, you hardly need foreign terrorists when you have homegrown menaces like the folks who ran the Peanut Corporation of America, whose factory was so filthy that their own inspections found salmonella on twelve occasions. But they kept shipping product, and the U.S. food system is so intertwined that they managed to poison nineteen thousand people in forty-three states and force the January 2009 recall of thousands of products ranging from Barefoot Contessa Espresso Dulce de Leche to Power Bone Canine Cookies.[28] Even without gross negligence, the industrial food chain is so complicated it's almost impossible to trace. In 2007, frozen pot pies sickened fifteen thousand people in forty-one states. But investigators were never able to pinpoint a single source. Was it the turkey? The peas? The carrots? A *New York Times* investigation found that ConAgra Foods, one of the planet's largest food suppliers, simply "could not pinpoint which of the more than 25 ingredients in its pies was carrying salmonella." It also found that other food giants "do not even know who is supplying their ingredients, let alone if those suppliers are screening the items for microbes." The solution? ConAgra put a sticker on the box of the sixty-nine-cent pies advising that "internal temperature needs to reach 165 degrees Fahrenheit as measured by a food thermometer in several spots." When staff members in the newspaper's test kitchen tried to follow the directions, they found that "some spots in the pies heated to only 140 degrees even as parts of the crust were burnt." Hungry?[29]

Our food supply, in other words—by far the most important thing on earth, at least if you're a human—is too big to fail. And it's failing.

Which is why it's useful to spend a day in Hardwick, Vermont.

When you think of Vermont—white church on the tidy green—you're not actually thinking of Hardwick, which in its days as the "Building Granite Center of the World" used to boast a dirty movie theater and a lot of bars. And those were the good times. In 2005 an enormous fire wrecked the historic block in the middle of town.

Likewise, when you think of compost, you may imagine a healthy-looking gardener spreading the loamy remains of her former vegetable soup on the raised beds where she'll grow next year's carrots. That's not Tom Gilbert. He's healthy looking enough, but he's standing in a dusty parking lot high on West Hill Road, overlooking town. "You're surrounded now by three decomposing carcasses," he says to me, pointing proudly to a trio of brown mounds. Tom Gilbert runs the Highfields Institute, which "brought livestock mortality composting to Vermont." On a dairy farm, 5 percent of the herd is likely to die each year, so knowing what to do with the remains is important. "You don't want to just haul it out in the field; that's a lot of blood and bone that will go to waste" when it could be improving the soil. So here's the recipe: an eighteen-inch base of wood chips, a six-inch layer of sawdust, a thin layer of corn silage, the carcass, and then a cap of silage. "There's a full-grown Holstein in there; I put him in two weeks ago," says Tom, who sticks a two-foot-long thermometer into the pile. "145 degrees. If you go in with a shovel you'll find nice clean bone. We'll leave it a while longer, and the skull and the pelvis will still be there, but now they'll be brittle enough that you're not going to pop a tire if you drive the tractor over it."[30]

Gilbert composts more than cows. In fact, he's pioneered a rural composting system that collects much of the food waste from the surrounding areas: schools, farms, restaurants. The truck drives a seventy-six-mile route; some of the stops are fifteen miles apart, which reduces the economies of scale. Even so, once Gilbert's crew has got the garbage up on the piles, where they can roll it with a backhoe every few days, it doesn't take long before it turns into fertilizer. "If you assume every yard of compost offsets a yard of synthetic nitrogen, our little operation here is saving the equivalent of fifty-four thousand gallons of gasoline a year," he says. If Vermont as a whole recycled all its food waste, it could compost twenty thousand acres of vegetable fields, which would be enough to grow most of the produce its citizens consume. And Vermont is a little place. Imagine New York City composting; it's comparatively easy to collect food waste when there are more people living on the West Side than in all of Vermont. The resulting fertilizer would be enough to make New Jersey the Garden State once more. "Soil is the frontier of where we need to be going," says Gilbert.

But forget New York for a while. We'll return to the cities and suburbs soon enough, but Hardwick is interesting on its own. Some of Gilbert's compost gets trucked about a mile down the road to the gardens where High Mowing Seeds grows its product. High Mowing is one of the country's biggest organic seed companies, which means it isn't all that big—a couple of million dollars a year in revenue. But it sure is beautiful. "People say it's hard to grow organic broccoli and cauliflower," says Tom Stearns, the ebullient proprietor. "We try to find the ones that really crank—like these," he says, pointing to specimens approximately the size of beachballs. "Organics need to be really vigorous in order to outcompete weeds. And they tend to need more root hairs, because their fertility is widely distributed instead of being intravenously injected. These guys here are perfect if you like radishes—I don't much like radishes. This is golden frill arugula, a new variety we're adding in 2010. Here's an Asian green, *hun si thai*—just eat, eat. We have two fields up here, and we keep them a mile apart to prevent crossing. We have zucchini in one and pumpkins in the other. Or else you get pumpkinis. Or maybe zuckins."

The High Mowing warehouse is down the hill—metal shelves are filled with the beginnings of a million meals. Quinoa. Spelt. Here's a big bag of Tom Thumb popcorn. Italian flat-leaf parsley. But not just flat-leaf. Double-curl. Also triple-curl. Two young women are hunched over a cutting board, examining onions. "We have a new favorite," one reports. "Rossa di Milano. It really stood out. It beat the Red Baron. High blocky shoulders." Stearns is gushing on about his business—the fast growth, the network of small growers around the country, the sterling germination rates—but I'm dizzy from the simple fertility. "That bag over there has thirty pounds of cuke seed," he says. "That's sixty acres." Orders come in by the hour. "People who used to get five or ten packets are suddenly getting twenty or thirty. The ten-by-ten garden is suddenly becoming twenty-by-thirty. People are trying to put some food up. I love high oil prices."

Lots of his seed goes a few miles farther down the road, to the town of Craftsbury, where Pete Johnson has been pioneering year-round farming in northern New England. Johnson built a solar greenhouse on campus as his senior project at Middlebury College and then started thinking bigger. By now he's figured out how to move his greenhouses on tracks, so he can cover and uncover different fields and as a result grow greens eleven months of the year without any extra heat, allowing him to run his CSA (community-supported agriculture farm) year-round. Say your family wants the spring share; you'd pay $748 for a weekly basket from February to June, and in mid-April—a tough time of year for local farming—you'd get maybe a half pound of mesclun, a bunch of parsley and of scallions, three pounds of carrots, some early radishes, two pounds of beets, two pounds of fingerling potatoes, half a pound of oyster mushrooms, a loaf of local bread, a half gallon of local cider, and a half pound of local feta cheese. You can add a meat share: a five-pound chicken, some pasture-raised hamburger, a couple of locally farmed trout, a pound of bacon cured without nitrates. By Johnson's calculation, it all comes to 20 percent less than buying the same stuff at a supermarket. But we're used to thinking of local food as more expensive. "Compared to what?" he asks. "Compared to the absolute junkiest food you can buy in a supermarket? It's too bad we think we can't afford the most important thing in the world."[31]

Other seed from High Mowing is dispatched a few miles in the other direction, to the headquarters of Vermont Soy. The folks there hand it out to four or five Vermont farmers, who in turn produce the beans that become tofu and soymilk in the small factory here. (Only half the space is used for tofu; the other half somehow turns milk whey into varnish for furniture.) The owner, Andrew Meyer, grew up on a local dairy farm, the kind of farm that's been going out of business for decades as the milk industry turns into a commodity business dominated by huge western dairies. So he understands the need for a more regional food economy. "I think Vermont hasn't even tapped its capacity for growing food. Someday the train will come back, and we'll be sending a refrigerated car once a week, right to the Chelsea Market. We've got two of the biggest markets in the world right nearby, Boston and New York."[32]

So: Compost. Seed. Greenhouses. Train cars to New York. Maybe you don't need Kip Cullers after all.

But before we dispense with industrial agriculture, we need to consider not just Vermont, not just New York, not just the Western world. We need an answer that works for the *whole* planet, much of which is by any standard much too poor.

The standard plan for how they're going to get less poor goes like this: Move off the farm. Stop having peasants, start having your own versions of Kip Cullers, and free everyone else to do something that makes real money. Here's Jeffrey Sachs, our best-known and perhaps best-hearted development economist, describing the "ladder of economic development," with its "progression of development that moves from subsistence agriculture toward light manufacturing and urbanization, and on to high tech services." You begin with peasants who "typically know to build their own houses, grow and cook food, tend to animals, and make their own clothing. They are therefore construction workers, veterinarians, and agronomists, and apparel manufacturers. They do it all, and their abilities are deeply impressive." But they are also "deeply inefficient," because "Adam Smith pointed out to us that specialization, where each of us learns just one of those skills, leads to a general improvement of everybody's well-being."[33]

To reap the fruits of that efficiency, you need density: modern economic growth, as he points out, is "accompanied first and foremost by urbanization." As each farmer uses more oil and chemicals to produce more food, "an economy needs fewer and fewer o they move to the city, "drawn by higher wages that turn reflect the higher productivity of work in densely settled urban areas." And so it is. Anyone who's visited China in recent years and watched people swarm off the farm and into the city knows what he means. And as this happens, he adds, "fixed social orders" like caste erode, unable to "withstand the sudden and dramatic burst of technological change that occur during modern economic growth." Fewer generations live under each roof; the age of marriage is delayed; there's "greater sexual freedom much less directly linked to child-rearing."[34] You produce, in other words, that thing we call modernity, the liberation from family, from community, and, at base, from the soil.

Earlier this decade humanity passed a remarkable threshold: there were finally more people living in the city than in the country. Will that migration continue? In the early days of our current economic downturn, as many as 30 million people in China moved *back* to the countryside because their factories had shut down. My guess is that it's the reverse migration that will continue, most of all because—without stable weather and cheap oil—we won't be able to keep emulating the Kip Cullers model much longer, not here and certainly not in the poor world.

We need to stop thinking of farming in abstract terms, as a "low rung on the ladder of economic development," and remember again what it involves: using water and sunshine to grow plants rooted in soils that can provide the nutrients people need. If the amounts of water and sun and soil nutrients change, so do the prospects for farming. We've already discussed water and sun, so for a moment let's consider soil. When you have virgin soil (say, a prairie that's been growing grass for millennia until someone invents the plow that can break it), that soil is full of nutrients. Plant a carrot seed, and up springs the carrot. When you eat the carrot, you eat those nutrients—that's the whole "food" concept. But it means those nutrients aren't in the soil any longer and need to be replaced. For a very long time, in places like China, people replenished the soil by returning their excrement to the fields—"night soil" it was called. That's how China kept growing food on the same fields for millennia. But of course that depended on people living near to those fields, so they could carry their dung there each day. When you move up the "economic ladder of development" and dispatch your family to Shaoxing to sew T-shirts, that's not possible. Instead you have to start using the increasingly barren soil as a matrix for artificial fertilizer. Soil becomes stuff that holds your plant upright while you pour oil on it.

Consider what's happened to Chinese soils in the last few decades as they've made their farming more "efficient" and "modern" and "mechanized," as they've freed up millions of people to move to the cities. The organic content of soils on the northeast China plain dropped from 9 percent early in the century down to 5 percent in the 1970s and 2 percent in the 1980s.[35] Only fossil fuel in the form of synthetic fertilizer keeps it—or the Great Plains, or the Punjab, or the other grain-growing belts of the planet—functioning. But it's precisely that artificial fertility that has allowed populations to keep growing. By some estimates, "every third, or certainly at least every fourth inhabitant of the Earth is now alive" owing to the process that uses natural gas to produce ammonia and hence nitrogen fertilizer.[36]

This seemed like one of those traps there was no way out of. Convert back to older forms of agriculture, and every third or fourth person would starve. Or maybe more. Here's Dennis Avery, an evangelist for high-tech farming, explaining the future should we stop relying on the current model: "Unless we starved *half* the humans, we'd displace the remaining wildlife" by cutting down all the forests to grow more forage for the cattle whose manure we'd require.[37] The idea that only giant agribusiness can keep the world from starving is deeply ingrained in us. It's one of the things we *know* about the world, though how we know it is not so clear. We just do.

So it's unsettling (but also the first unambiguously good news this book has to offer) to learn that serious people have begun to rethink small-scale agriculture, perhaps just in time to help us deal with the strains of our new planet. In the last ten years academics and researchers have begun figuring out what some farmers have known for a long time: it's possible to produce lots of food on relatively small farms with little or nothing in the way of synthetic fertilizer or chemicals.

Some of these serious people are historians. In her account of wartime Britain, Juliet Gardiner delivers a statistical account at least as amazing as America's tale of converting Detroit to building bombers. In the course of the war years, Britain managed to increase food production 91 percent. Small gardens—*allotments,* the English call them—sprung up everywhere. For example, the wife of the keeper of coins and medals at the British Museum planted rows of beans, peas, onions, and lettuces at the museum's entrance. Almost seven thousand "pig clubs" sprung up throughout the country, with swine being kept in, among other places, "the (drained) swimming pool of the Ladies Carlton Club in Pall Mall."[38]

Most of the serious people are agronomists, however, who have begun to think more closely about the assumptions underlying our ingrained view that big is better. Take the question of yield, for instance. A monomaniac like Kip Cullers can produce huge quantities of soybeans or corn, and even the average industrial farmer can produce more of these crops on an acre than most small farmers. But what if you can grow two things at the same time?

I remember my confusion the first time I walked through a smallish cornfield north of Beijing. Hidden between the rows of corn were rows of green beans; the technical name for this is *intercropping,* and it's as old as farming. But we've forgotten how because if you've got a huge tractor to drive, monocultures are the only way to go; Kip Cullers's combine would squash the beans every time he went out to cultivate the corn. Here's the Indian agronomist Vandana Shiva: "The Mayan peasants in the Mexican state of Chiapas are characterized as unproductive because they produce only two tons of corn per acre. However, the overall food output is 20 tons per acre. In the terraced fields of the high Himalayas, women peasants grown jhangora (barnyard millet), marsha (amaranth), tur (pigeon pea), urad (black gram), gahat (horse gram), soybean, bhat (glysine soya), rayans (rice bean), swanta (cowpea), and kodo (finger millet) in mixtures and rotations. The total output, even in bad years, is six times more than industrially farmed rice monocultures."[39]

It's not just making better use of space. Small farmers often make better use of time, too, figuring out how to plant more crops in a year. Because by definition small farmers can spend more hours in each of their fields, they know them better; they're farming across seven acres, not seven counties, so they can plan on how to get the most out of every dip and ridge. And because they usually own the land they plow, peasant farmers work harder and more creatively than hired hands—there's no more conservative idea than this, even if it has radical implications. World Bank economists, for instance, "now accept that redistribution of land to small farmers would lead to greater overall productivity." Reexamining data from the last few decades, economists have found that small farms are more productive in

Africa, Asia, and Latin America.[40] That's especially amazing, since the best land in developing countries usually goes for export crops, with locals "often reduced to growing their own food on rocky outcrops or steep slopes."[41] In North America, center of the mechanized megafarm, the U.S. Department of Agriculture reports that according to its latest census smaller farms produce far more food per acre, whether you measure that output in tons, calories, or dollars.[42]

Better yet, those small farms are capable of getting far more productive with each passing season, because they can take advantage of new information, new science, new technologies. We're used to thinking of ag science as synonymous with Monsanto, and new technology meaning that $185,000 GPS-guided tractor. But there's much more interesting work than that under way, in unlikely places around the globe. It's producing a hybrid, a new/old agriculture that dispenses with many of the expensive inputs and relies instead on information and organization. Jules Pretty has seen more of this than anyone else. An agricultural economist at the University of Sussex in England, he's led a study of 286 "alternative agriculture" projects involving 12 million farmers all over the world.

In Indonesia, a million farmers have gone through informal "farmer schools," learning from their neighbors new ways to control bugs by encouraging their natural pests. "The yields of rice they get are roughly the same," says Pretty, "and of course they're saving lots of money on pesticides. But the big boost comes from the fact that once you stop using pesticides you can grow fish in the rice paddies. The fish help to circulate nutrients, they eat a lot of pests themselves—and you've got seven hundred kilograms of fish per hectare per year to eat or to sell. That's a lot of protein."

Or look at the maize fields of East Africa, where farmers have made big inroads with agroforestry in recent years, using trees and shrubs that fix nitrogen in the soil and produce lots of leaves that you can till into the soil to improve fertility. "You can't grow corn every year on these systems—you have to let it lie fallow some of the time—but you get such a big boost in yields that it's worth doing," Pretty explains. "And of course you get wood from the trees and shrubs themselves, so you're not having to walk forever to get firewood. Many rural people in Africa grow trees primarily to raise money for school fees for their kids," which means the cycle of knowledge should continue to grow.

These new schemes can upend the conventional wisdom. "If you pick up a soils textbook," says Pretty, "it will tell you that soils take geological time to create. But I've seen places in Honduras where farmers have created a meter of soil in ten years—where the ground has gone from played-out desert to springy black loam." The key in many of these farms was a new crop, velvet-bean, which farmers learned to grow between the rows of corn. "It fixes vast amounts of nitrogen, and you use the plant as a so-called green manure. You put it on the soil, it rots down, and it effectively creates a new soil on top of the old one. Many fields will triple or quadruple their yields—and all you have to do is learn to manage a mixed crop." This is a hard lesson for large agribusiness farms used to the efficiency of growing just one thing, but relatively easy for peasant families who are always there on the ground.[43]

The new agriculture often works best when it combines fresh knowledge with older wisdom. In the cornfields of East Africa, British scientists working with their Kenyan counterparts discovered a new class of so-called semiochemicals that plants produce to attract or repel pests. They worked out a planting system designed to thwart the corn borer, one of those pests expected to increase its range as the temperature warms: you grow grasses on the edge of the field that attract the parasites that feed on the borer, and next to them other grasses that simply repel the borer (and then, of course, you feed the grass to the livestock). "Local knowledge alone couldn't have figured that out, because you can't see or smell those semiochemicals," says Pretty. On the other hand, the official agronomic advice for forty years had made the problem much worse—creating big monocultures with modern varieties, and then soaking them in pesticide and fertilizers. Now four thousand farmers in Kenya are using the so-called push-pull system, *vutu sukumu* in Swahili. Depending on how you think about it, they're considerably more modern than old Kip Cullers.

To see this kind of ingenuity at work is to understand its power. I've reclined under a palm tree in Bangladesh where a hundred species of fruit and vegetable grew in a single acre: the farm featured guava, lemon, pomegranate, coconut, betel nut, mango, jackfruit, apple, lichee, chestnut, date, fig, and bamboo trees, as well as squash, okra, eggplant, zucchini, blackberry, bay leaf, cardamom, cinnamon, and sugarcane plants, not to mention dozens of herbs, far more flowers, and a flock of ducklings. A chicken coop produced not just eggs and meat but waste that fed a fishpond, which in turn produced thousands of pounds of protein annually, and a healthy crop of water hyacinths that were harvested to feed a small herd of cows, whose dung in turn fired a biogas cooking system. The guy who was showing me around summed it up like this: "food is everywhere, and in twelve hours it will double." More to the point, in the new world we're creating it's like having a hundred small banks instead of one big one. If it gets too dry for squash, the coconut will probably see you through. You're not putting all eggs in your basket—there's some eggs, but there's some eggplant, too.

• • •

As I said before, it's hard for us to see this. We've trained ourselves to think that "producing enough food for a growing world" means "big tractors." In the fall of 2008, the United Kingdom's former chief scientist, Sir David King, blamed "anti-scientific attitudes" among Western nongovernmental organizations for "holding back" a new "green revolution" across Africa. Organic farming across the continent would, he insisted, have "devastating consequences." But a month later, the United Nations Environment Programme issued a report showing that yields across Africa "doubled or more than doubled where organic or near-organic practices had been used." In East Africa, harvests jumped 128 percent. Not only were harvests better, but the organic soils were retaining water and resisting drought. "Saving money on fertilizers and pesticides help farmers afford better seeds" too.

Some of the practices were simply traditional farming, while others drew on Western innovations like double-dug beds. Henry Murage, a small farmer on the western slopes of Mount Kenya, spent five months in England, studying with experts at an experimental farm in the Midlands. When he returned to Africa, he persuaded three hundred of his neighbors to adopt at least a few of his practices—during the last devastating drought to hit the area, they were the ones who fared best.[44] It's harder work at first; anyone who has double-dug their own beds can remember the knot between the shoulders. But once the work is finished, in the words of one agronomist, "little has to be done for the next two or three years." Pretty reports that in one review of Kenyan organic farming in twenty-six communities, "three-quarters of participating households were now free from hunger during the year, and the proportion having to buy vegetables has fallen from 85 to 11 percent." He describes a woman, Joyce Odari, with twelve of the raised beds, so productive that she employs four young men from the village to tend them. "The money now comes looking for me," she says.[45]

The systems get better over time, not simply because the soil improves but also because farmers stop relying on the rote advice of chemical companies and start paying attention to their fields. In Malawi, tiny fish ponds that recycle waste from the rest of the farm yielded on average about 800 kilograms of fish when they were begun in the 1990s; half a decade later that figure was 1,500 kilograms. Instead of playing themselves out, the way our industrial soils have, these farms were revving up. And as the farmers talk among themselves, new ideas spread quickly. In Madagascar, rice farmers worked with European experts to figure out ways to increase yields. They transplanted seedlings weeks earlier, spaced them farther apart, and kept their paddies unflooded during most of the growing season. That meant they had to weed more, but it also increased yields fourfold to sixfold. "The proof that it works comes from the number of farmers using it—an estimated 20,000 farmers" have adopted the full system, according to Jules Pretty, and another 100,000 are experimenting with it. Now word has spread to China, Indonesia, the Philippines, Cambodia, Nepal, the Ivory Coast, Sri Lanka, Bangladesh.[46]

And even to the grain belts of the United States, where the same kind of transitions are under way. Klaas Martens grew up on his parents' farm in upstate New York. He went away to Cornell, the state's main ag school, in the early 1970s, and he returned with a firm belief in the whole arsenal of chemicals that constituted "modern" agriculture. "I had every trick in the book," he says. "And at first we saw yields go up phenomenally." He kept close records, though, and in a few years "we saw changes in our soil, we saw weeds we'd never seen before, new diseases. Pretty soon we needed two herbicides, then a combination of three.

Every time we thought we'd figured out how to make things work, nature would make it not work." His wife was no earth mother either; she was in charge of figuring out the pest-spraying program for Concord grapes, a dominant crop on the western edge of New York. But both of them were noticing health problems, especially during spray season, and with children on the way they decided to switch to the then-infant field of organics. "Experts—the people I trusted—said it might work on a small scale, in a garden maybe. But not on a farm. Pests would descend from every direction," he recalls.[47]

Instead, what's descended have been other farmers, who over the years have watched the Martenses' 1,400-acre farm—mostly grain, especially livestock feed—prosper without chemicals. But it's not what they *don't* use that defines their farm, Mary-Howell Martens insists. It's what they do use: "We see it as just as much precision farming as any other kind. But we substitute observation, management, planning, and education for purchased inputs." Working with a Cornell agronomist who's been studying their fields, they've figured out that they can avoid the root rot that afflicts dry beans if they plant buckwheat the year before in the same soil. They don't need the genetically modified BT corn that many of their neighbors use to deal with corn borers. "We find that we don't have a borer problem if we have a healthy rotation," Klaas Martens reports.

At their farm and at the grain mill they rebuilt to handle a growing volume of organic grain, they now work with dozens of other farmers making the switch to low-input growing. "The first couple of years there usually is a yield hit. You don't have the rotations in shape yet, or the right weed control equipment, or quite the right timing," explains Klaas. "But after they've been doing it for a few years, people's yields come back up, and people stop feeling so panicked." In recent dry years their harvests have shone—their soils, richer in organic matter, hold water better. "In wetter years like this one [2008] it's a little rough. When the rains came, the weeds came on really strong. Still, we've got phenomenal corns and soybeans this year."

So I'm willing to listen when Mary-Howell says, "Can organics feed the world? If the world changes its diet. We need more crops rotating, a more diverse diet, not all based on high fructose corn syrup."

"We'd definitely need a mosaic," adds Klaas. "Diversity is everything. It goes all the way down to a handful of soil—which has more species, more biodiversity, than a whole square mile above ground. And the way we farm is killing a lot of that right off."

It's a gospel they keep spreading. The number of farms has actually begun increasing in many parts of the United States in the last few years, and all the growth is coming in just this style of agriculture. Across New England, where farms have been dying for 150 years, the number of farms grew from 28,000 to 33,000 between 2002 and 2007, even as the average farm size fell from 142 to 122 acres.[48] Nationally, 300,000 new farms sprung up this decade.[49] I was talking to the Martenses in the fall of 2008, and they had just been to speak at a conference outside Albany, of new growers who managed plots from an acre to twenty-five acres, many of them carved from failed conventional farms. "There were fifty farmers there, and almost all of them had grown up in cities," reports Klaas. "The numbers are starting to shift back."

But if our farming is *really* going to shift back—in time to let us deal with the new conditions we've unleashed here on planet Eaarth—it will take dramatic change. For one thing, all the kinds of innovative farming I've described share one feature: they require more people than conventional farming. Not one

Kip Cullers in his mighty machine, but lots and lots of folks down on the ground. For a hundred years we've substituted oil for people, which is why we have more prisoners than farmers in the United States; now we need to go the other way. One British study found organic farms need 4.3 workers per 250 acres, double the number on a conventional field.[50] They're picking the bugs that the pesticides used to kill and turning the compost. In Burlington, Vermont's chief city, the Intervale farms supply about 8 percent of the produce residents consume from just 120 acres of soil, but in midseason there are *fifty people* working those 120 acres. Conversely, plantation soybean production for biofuels in Brazil uses one worker in the area that eleven subsistence farmers once tilled—which is why, given our current economic calculus, those plantations are expanding and those peasants are relocating to cardboard boxes on the edge of Rio de Janeiro and São Paulo.[51]

Putting more people on the land would take getting used to, but it's possible. In the United States, local land trusts, which used to concentrate on preserving unspoiled scenic views, increasingly save land precisely to turn it over to small farmers. In Brazil, where landless peasants have started moving out of the cardboard boxes and squatting on industrial plantations, they are finding a steadily more pleasant welcome. "They sell their produce in the marketplace of the local towns, and buy their supplies from local merchants," explains one researcher. Towns with such "land occupations" find themselves better off than their neighbors; some mayors are actually petitioning the Landless Workers Movement to come carry out occupations in their areas.[52]

The biggest shifts required will be political. There are "farms" in the United States with hundreds of thousands of swine, producing more sewage each day than big cities. The animals are miserable, the pollution is intense, and it's all utterly unsustainable—by some estimates, as much as half of global warming gases can be tied to the livestock industry, with its huge demands on our grain crops. But it makes a few big corporations rich, and it keeps the price of food incredibly low, and that bargain—enshrined in one federal farm bill after another—has prevailed for two generations. We'd need to challenge the power of those companies, and we'd need to be willing to pay our neighbors enough to grow our food so that they could lead decent lives.

It probably shouldn't cost *much* more. Eliminating all the middlemen that take most of the agricultural dollar would keep prices affordable. By some estimates, seventy-five cents of every dollar spent on supermarket food covers the cost of advertising, packaging, long-distance transport, and storage; at a farmers' market, by contrast, 95 percent of the price goes to the farmer growing the food.[53] For poor people, the price is particularly right; since inner-city supermarkets typically charge a third more than suburban ones, and since the produce on the bodega shelf usually defines wilted, urban farmers' markets allow what one story called "ready access to wholesome and cheap food." When such a market is nearby, the consumption of fruit and vegetables increases,[54] and recent immigrants are often the most enthusiastic customers (perhaps because they can remember what actual food tastes like). They also save money because the food's fresher: a local vegetable "might last a week in the fridge, whereas one that's traveled might last only two days, since it's aged," says the Tufts University analyst Hugh Joseph.[55] Since one out of four fruits and vegetables never makes it to the table because it spoils, that adds up.[56]

We'd need to eat a little differently, too. I was born in 1960, when the average citizen of the developed world ate 116 pounds of meat annually. That's grown to 187 pounds in just fifty years, which means that at the moment Americans eat 200 pounds of grain directly, and 1,800 pounds that's been run through an animal first. By contrast, the average Chinese eats 850 pounds of grain with his chopsticks, and only 154 pounds indirectly, via cow or pig. It takes eleven times as much fossil fuel to raise a pound of animal protein as a pound of plant protein,[57] which means we'd be wise either to turn vegetarian or to take a Chinese cooking lesson. The Chinese use meat almost as a condiment, adding flavor, instead of employing the Big Honking Slab culinary technique favored by Americans.

Spreading all those changes fast enough to make a real difference will take a change of heart. We'd need to view food as something important again, worth taking the time to seek out and to cook. But you can see it starting to happen: local farmers' markets are the fastest-growing part of the food economy, with sales up by 10 to 15 percent a year, and the number of markets doubling and then doubling again in the last decade. If governments invested modest sums in infrastructure, and reduced the regulations that make it hard to sell a side of beef to a neighbor, the expansion would accelerate. If the government simply stopped subsidizing commodity agriculture, that would likely be enough to tip the balance: 61 percent of the billions in farm subsidies that we pay each year go to the largest 10 percent of American farmers. Those sums skew the economics of farming against small local producers—the ones who supply food, as opposed to corn syrup, and satisfying jobs, as opposed to hopeless hard labor. Here in verdant hippie Vermont, for example, most of our few remaining dairies have gotten so big that they employ mostly illegal Mexican immigrants, the only people willing to work for the wages, and under the conditions, that the dairies require if they're going to compete with the even bigger ones out of state.

(And even so they're losing money. A recent analysis found that the state's dairy farmers were managing to lose more than a quarter billion dollars annually.)[58] You rarely see the migrants, though there are at least two thousand in the Champlain Valley; they live in the shadows, often barely leaving their makeshift farm dorms for fear of being arrested. Contrast that with the forty applicants a week looking for jobs at High Mowing Seeds, jobs that pay forty thousand dollars a year and come with medical benefits, not to mention as much produce from the test gardens as you can cart home.

If millions of people are going to work at least part-time as farmers again, they won't all be in rural America. Much of the best land in the country—flat, fertile soils within easy reach of cities—has been converted to suburban tract housing. But there's a lot of acreage left around each dwelling; an analysis of available land in American suburbs recently suggested that they could, on average, "realistically provide around 50 percent" of the food they need, acting "as a localized buffer against disruptions, and providing a high percentage of vitamins, minerals, flavor, and culturally-important foods."[59] The number of American gardens grew 10 percent in 2008 and was expected to increase faster still in 2009; Burpee Seeds was one of the few companies prospering in the recession, reporting a sales rise of 40 percent between 2007 and 2008.

Many of these new gardens lie in our biggest urban areas. Los Angeles reports a waiting list for the ten-foot-by-thirty-foot plots it leases at seventy sites around the sprawling city.[60] New Jersey? It would take 115,000 additional acres to feed everyone in the state a healthy vegetarian diet, the equivalent of 3.4 million two-hundred-square-foot gardens.[61] That's ten beds, each four by five feet—which could fit in your yard, over there next to the swing set. New York? The city itself, by one study, could produce between 10 and 20 percent of its produce.[62] (Shanghai, by comparison, employs 270,000 people in its urban farming industry.)[63] And there's abandoned farmland surrounding the city—huge swaths of what we call "upstate New York" we once called America's "grain belt." Food processing is returning to the city too; a recent *New York Times* survey of Brooklyn found entrepreneurs making everything from cheese to pickles to beer.[64]

This renewal is under way across much of the country. If you want a sense of what's possible, pull out a local map and see how many roads are named for the gristmill that once lay along them. In western Massachusetts, one bakery that wanted local wheat found that none of the farmers in the area had any idea anymore what varieties to plant, and they lacked the facilities to clean, mill, and store the flour, so the owners persuaded a hundred of their neighbors to plant small patches in their yards. When New Mexico officials started a wheat-growing project in 1994, they found that the crop hadn't grown there for half a century; now local growers sell 350,000 pounds of flour a year.[65] Too expensive? In 2008, when food prices soared, our local Vermont wheat farmer Ben Gleason kept his price at fifty-nine cents a pound, even when the price of King Arthur flour was twice that. His production costs hadn't gone up much; it didn't seem right to gouge his neighbors.

I'm not arguing for local food because it tastes better, or because it's better for you. (It does, and it is. There's really not much debate.) I'm arguing that we have no choice—that the new Eaarth has much less margin than the planet we grew up on, and hence we're going to need to take advantage of opportunities we've passed by before. In a world more prone to drought and flood, we need the resilience that comes with three dozen different crops in one field, not a vast ocean of corn or soybeans. In a world where warmth spreads pests more efficiently, we need the resilience of many local varieties and breeds; in the past century five thousand domestic breeds of animals and birds went extinct, and each time our danger increased a little.[66] And in a world with less oil, we need the kind of small mixed farms that can provide their own fertilizer, build their own soil.

The vast conformity of our agriculture puts us at risk—there are no more firewalls than there were in our financial system. "Eager to find a super-species," writes Michael Shuman, "growers around the world are relying on a dwindling number of carefully designed seed species. These monocultures are genetically 'stable,' which means that each seed grows into fruits and vegetables similar in size, color, and taste."[67] And even then the grocers are picky. The British journalist Felicity Lawrence found recently that for "every 10 tons of carrots harvested, only 3 tons pass muster. In Kenya, some 35 per cent of the bean crop grown for export is thrown away because it does not meet the supermarket specification that beans must be straight, of an exact diameter and length, and cosmetically perfect."[68] By contrast, Shuman says, if we let plants evolve in particular places according to the local temperature, soil, and pests, "thousands of subtle genetic variations inevitably emerge with thousands of unique defenses."[69] In the summer of 2009, for instance, late blight wiped out the tomatoes in many vegetable gardens across the northeast United States—apparently because so many new gardeners bought their starter sets from Wal-Mart or Home Depot, revealing that they all came from an industrial-scale operation in the South that "transferred their pathogens like tiny Trojan horses into backyard community gardens."[70] Instead we need the local differences in variety, planting technique, and timing that have been designed over centuries of trial and error "to produce the most stable and reliable yield possible under the circumstances," writes James Scott, an authority on peasant agriculture. Typically, the small farmer seeks to avoid the failure "that will ruin him rather than attempting a big but risky killing."[71] You just lived through the subprime mortgage debacle. Want to double down on your dinner? Just *look* at a small farm. In the United States, on average, 17 percent of a small farm's area remains in woodland, compared to 5 percent on large farms. Small farms keep twice as much land in "soil-improving uses," like cover crops. There's margin there for the preservation "of hundreds if not thousands of wild and cultivated species," in the woodlot and in the fallow field and in the fishpond and the back garden.[72]

Does it *really* matter? Hurricane Mitch hit Honduras and Nicaragua in 1998. Since it didn't hit the United States, the storm quickly faded from our memories, but the Central Americans will never forget—Mitch killed more people than any Atlantic hurricane on record except the unnamed great storm of 1780. A huge slow-moving tempest, it dropped three feet of rain in many places. In Choluteca, Honduras, 212 days' worth of rain fell in twenty-four hours, and the local river flooded to six times its normal width. Honduran president Carlos Roberto Flores said the storm destroyed fifty years of progress in his country; it wiped out nearly every bridge and secondary road. "The damage was so great that existing maps were rendered obsolete." And of course farmers took it on the chin: 70 percent of crops were wiped out.[73] But here's the thing: after the storm, a study of 1,800 farms found that small farmers using sustainable practices suffered far less damage than their conventional neighbors. Diversified plots had 20 to 40 percent more topsoil, greater soil moisture, less erosion. They suffered far fewer economic losses than their conventional farm neighbors.[74] They didn't *enjoy* the violent new weather we've created. But they survived it.

Very hardheaded. Entirely practical. But if you want just a

touch of the sentimental, a recent study of sixty-three inner-city community gardens found that they were supplying more than produce. Donna Armstrong also found that "they changed local residents' attitude to their own neighborhood, resulting in improved care for properties, reduced littering, and increased pride in the locality." Community gardens, she found, "promoted social cohesion, and encouraged people to work cooperatively on a range of local needs, such as shared child-care." They brought people out of themselves and into the community—they made those tough neighborhoods more resilient. Four in five of the people who worked the gardens said their mental health improved.[75]

Want me to lay it on thick? A few years ago, PBS broadcast a "reality series" called *Frontier House,* which required several families to remove themselves to a re-created Montana circa 1880. The usual laff riot ensued—one family snuck off the ranch to trade their home-baked goods for some meat and the chance to watch a few minutes of MTV. The weird and unpredictable stuff happened when the show ended, and the six adults and seven children moved back to the suburbs. A majority of the grown-ups, and all the kids, said they preferred their lives on the frontier. One man moved back to his old cabin in Montana and took up life as a ranch hand. One of the kids grew seriously depressed, and they all reported that they missed having chores, taking care of the animals, and actually seeing their parents. Forget the guys with the bar-coded overalls that link to your Facebook page; "in five months in 1883 I got more satisfaction, more accomplishment, more appreciation than I did in my entire career beforehand," said one of the participants. One woman was building a five-thousand-square-foot home back in the "real world," where houses the size of junior high schools dot suburban ridges. Suddenly, she said, it felt a little . . . big.[76]

• • •

So there will be dinner, if we're resourceful and clever, and if more of us are willing to do the work of farming, and if we build the kind of community institutions that make us more resilient, less vulnerable. It won't be easy; as flood, drought, and pests spread, we'll be pressed to keep up. And it won't work everyplace; even the best double-dug community-backed garden still needs water. I don't know what Las Vegas will do. But many places may still produce enough calories.

Let's now move on to light and heat and the other good things that fossil fuels have brought us. How is it possible to imagine them on a globe that's running out of oil and running out of atmosphere? The answer is much the same as with food: local and dispersed works better than centralized, at least in a chaotic world.

To begin with, by this point it should be pretty clear that, even more than food, fossil fuels define "too big to fail." By burning every gallon of oil and cubic meter of gas and ton of coal we could find, we've managed to end the climatic stability that's marked human civilization. We've also managed to bet our entire economy on the belief that these supplies will last forever, a bet we're now in the process of losing. So far that failure has looked like: four-dollar-a-gallon gasoline and the economic meltdown it helped incite. It's looked like Russia temporarily shutting off the gas to Ukraine—and in the process to much of western Europe. In the southeast United States, in the wake of Hurricanes Gustav and Ike, a newspaper editor captured what it looked like in a headline: "Drivers Follow Tanker Trucks Like Groupies."[77] (The storms had knocked the main pipeline to the area out of service for a few days, which was all it took to reduce drivers to "camping out" for four hours at a time by pumps, waiting for the next delivery to arrive.) A Pentagon study concluded recently that you don't even need a storm—that in a single night, without ever leaving Louisiana, "a few saboteurs could cut off three-quarters of the natural gas supplies to the eastern United States for more than a year."[78] Even a rumor or two could do the trick. The oil analyst Matthew Simmons predicted recently that "a run on the gasoline bank" could collapse the oil sector of the economy in thirty days, noting that if "car owners decided overnight to fill up all 220 million cars in this country, it could drain most of the gasoline inventory dry."[79] You can't overstate how giant and interconnected and top-heavy this system has become. On a typical day, when about 43 million tons of goods are carried on U.S. transportation networks, *a third* of that tonnage is coal and oil.[80]

So here's the needle we need to thread: in the space of just a few years we've got to switch away from fossil fuel.

First: if we are serious about returning the atmosphere to 350 parts per million carbon dixiode—which is what we must do to stabilize the planet even at its current state of disruption—we need to cut our fossil fuel use by a factor of twenty over the next few decades.[81] As the British author George Monbiot told a group of protesters camped on a runway at Heathrow Airport in 2007, "We're not talking any more about measures which require a little bit of tweaking here and there."[82]

Second: it would be nice to replace at least some of that fossil fuel with something else, so that we're not returned entirely to a world of manual labor, where muscle power provides almost all the energy. Remember, a barrel of oil equals about eleven years of manual labor, and the average American uses the equivalent of sixty barrels a year.[83]

Third: there is no easy way out. I've already described why we're not going to be able simply to build a lot of nuclear plants—expense, mainly—or just convert our whole corn crop into gasoline (we need it to eat). But we're not going to toss up some solar panels and windmills and carry on as before either. We had our magic fuel: coal and gas and oil were amazing stuff, incredibly concentrated, buried just a little ways beneath the ground, often under such pressure that they'd literally spurt to the surface. Sun and wind are the opposite: soft power, diffused over wide areas, not always available. So no silver bullet—but maybe enough silver buckshot if we gather it carefully.

Job one, on almost anyone's list, is conservation, because the less power we use, the less sun and wind we'll need to capture. The numbers are huge: the consulting firm McKinsey estimated in 2008 that existing technologies could cut world energy demand 20 percent by 2020. To understand why, tour a typical home in your neighborhood. Maybe there's no insulating blanket around the water heater, which is turned so high that you need to cut the temperature with cold water when you turn on the shower. Video-game consoles can use as much juice as two refrigerators when they're left on, which is likely to happen because manufacturers "ship them with the auto power-down disabled."[84] This list goes on a long time—we are energy wastrels. Which in a sense is good news: the first round of cuts will be easy, like losing weight by cutting your hair. The energy analyst Amory Lovins recently calculated that Americans could, relatively cheaply, save half the oil and three-fourths of the electricity they use.[85] Does that sound far-fetched? DuPont, which is not quite Ben and Jerry's in its commitment to social progress, has managed to cut energy use 6 percent a year since 2000 by focusing on efficiency. That means it's using half what it used to.[86]

Since our homes are scattered, local action has the best hope of reaching most of them. Last winter, a man in our town of five hundred decided to train a few neighbors to conduct home energy audits. Twenty-two people showed up at town hall for training, and they then spread out around the community, armed with compact fluorescent lightbulbs, programmable thermostats, and insulation for hot water pipes. It worked—as local action often does. Washington never did sign on to the Kyoto agreement, but more than nine hundred American cities pledged to meet its targets, and some of them have actually succeeded. Portland, Oregon, managed to cut per capita carbon emissions by more than 10 percent, even as the local economy boomed. Leaders looked around their city and figured out what made sense: cheap bus passes for city employees, 750 miles of bike paths.[87] But it's not all government. SHIFT, a local group devoted to "demonstrating that cycling is liberating, fun, and empowering," hands out doughnuts and coffee to bicyclists the last Friday of each month as they cross the city's bridges. (If you're moving, they'll also assemble a team of cyclists with bike trailers to get the job done. "We do this for friendship and good times, not for money," though [local] beer and a meal at the end of the day are recommended.) Another local group specializes in "intersection repair"—cobblestoning intersections, painting them bright colors, building sculptures on the corner. Cars slow, foot traffic increases, neighborhoods start to cohere instead of sprawl.

Still, that's the soft stuff. Conservation, community, bikes. What about the real nuts-and-bolts engineering?

For generations, just as we've assumed that food must come from a distance, we've assumed that energy will be produced in a few centralized locations and then transported in our direction. The scale of those generating stations is immense. I remember standing in Con Ed's Astoria generation station in New York City one day and gazing in wonder, not just at the throbbing power of the boilers but also at the fact that the shift supervisor needed a bicycle to keep track of his domain. And in fact that plant was small. America's largest, the Scherer Plant operated by Southern Company in Lamar County, Georgia, needs three coal trains a day to keep it supplied. Each train is two miles long; the writer John McPhee described the scene when they pull into the plant. Car after car pulled atop a trestle, where compressed air swung their bottoms open, and as the coal dropped out onto the great pile that feeds the plant, "the uncontrollable dust far below had the look of an occurring disaster, the spreading clouds dark and flat as if they were derived from incendiary bombs."[88]

Given the need to replace such monstrosities, it's natural to think big. Even for renewable energy, size makes a certain kind of sense: though the sun shines on every human, and the wind rustles every blade of grass, it shines and rustles a good deal harder depending on your location. If you're going to build big concentrated solar power arrays, start in the desert Southwest (or, for Europe, in North Africa and the Iberian Peninsula); the biggest wind farms need the steady gusts of the Midwest, or the reliable onshore breezes on either coast of the Atlantic. We've recently begun to build some of this infrastructure, including the transmission lines necessary to connect them to places where most people live.

In one sense, though, constructing a huge national green energy system is sort of like buying organic food at the supermarket; it's an improvement, in that the fields where it's grown aren't soaked in pesticides, but that produce is still traveling an enormous distance along vulnerable supply lines. And instead of building stronger local communities, the money you spend buying it just builds the bank accounts of a few huge firms. With food, people are starting to understand the virtues of going not just green but *local*—and the same thing might be happening with energy. For two decades some farmers have built CSAs, or community-supported agriculture operations, where members pay an annual fee for a share of the produce. Now advocates like Greg Pahl are talking about CSE, or community-supported energy, and pointing at examples like the wind power associations and cooperatives that have built thriving facilities across Germany, Denmark, Holland, Sweden, and Canada.[89]

It makes financial sense to generate your power close to home; most communities spend 10 percent of their money paying for fuel, and almost all of it disappears, off to Saudi Arabia or Exxon. But is there really enough energy out there for most places to make their own? Can we really have the equivalent of farmers' markets in electrons? It sounds unlikely, but think a little harder. Say you cut your energy use by a quarter—perhaps by turning off the video-game console once Grand Theft Auto was done for the day. Then say you put some solar panels on your roof, the energy equivalent of a backyard garden. They probably couldn't supply you with all your juice, but some of it. Now the task of providing the rest locally has begun to look a little less daunting.

In fact, in 2008 the Institute for Local Self-Reliance published a series of studies that showed that half of all American states could meet their energy needs entirely within their borders, "and the vast majority could meet a significant percentage." Wind turbines and rooftop solar panels could provide 81 percent of New York's power, for instance, and almost two-thirds of Ohio's. But since North Dakota could provide *14,300 percent* of its power needs, almost entirely from windmills, you might think the most logical course would be to simply concentrate on building turbines near Fargo and then ship the energy to Akron and Dayton; after all, it's 30 percent cheaper to spin those blades in the Dakotas than in Ohio. But it turns out the math is more complicated. For one thing, the new transmission lines necessary to carry that wind power east would run at least $100 billion.[90] The institute's analysis found that once you factor in the cost of building the transmission lines, and subtract for the amount of electricity that's lost by sending it long-distance, the cost to Ohio consumers would be "fifteen percent higher than local generation with minimal transmission upgrades." You need 19 percent higher wind speeds to offset the cost of a five-hundred-mile transmission line; that's about the longest distance, they found, that shipping wind power really makes sense. Meanwhile, the local wind farm would employ almost twice as many people, and the profits from financing it would go to local banks instead of giant money-center operations.[91] Building those big new power lines costs as much as $10 million a mile, according to Massachusetts energy secretary Ian Bowles, and "there are better and cheaper ways to get more clean power flowing to the big cities," since every region has real potential for clean energy: hydropower in the Northwest, offshore wind in the East, solar energy in the Southwest. "In the Southeast, biomass from forests may one day be a major source of sustainable power. In each area developing these power sources would be cheaper than piping in clean energy from thousands of miles away."[92]

By the summer of 2009, in fact, the main force holding back distributed power—the microgrid—was the simple reluctance of big utilities to surrender their monopolies. With their political sway, they were backing congressional bills that, in the words of business journalist Anya Kamenetz, would "subsidize construction of multi-billion-dollar far-flung supersize solar and wind farms covering millions of acres, all connected via outside transmission lines." An array of research demonstrated that the faster, cheaper way to build renewables was to find a few rooftops on every block and put up panels, but "the utilities have been so adamant about thwarting these programs," says Jim Harvey, the founder of the Alliance for Responsible Energy. But don't take his word for it—ask the industry itself. Here's Ed Legge of the Edison Electric Institute, the chief lobbyist of the nation's utilities: "We're probably not going to be in favor of anything that shrinks our business. All investor-owned utilities are built on the central-generation model that Thomas Edison came up with. You have a big power plant. . . . Distributed generation is taking that out of the picture—it's local."[93]

Powerful as the utilities are, they may not be able to forever defy physics and economics. In July 2009, for instance, T. Boone Pickens pulled the plug on the world's biggest wind farm he'd planned for the Texas Panhandle because the transmission lines were too expensive; he planned instead a series of smaller installations closer to major cities. On the East Coast, by contrast, plans were still moving ahead for a series of offshore wind farms. "It just makes more sense to harness offshore wind right next to these big demand centers," said Jim Gordon, the CEO of Cape Winds.[94] The engineers call it *distributed generation*, producing energy where it's needed instead of ferrying it great distances. Coal plants need to be out of sight—they're dirty, poisoning the air around them. But more and more companies are installing "micropower" plants, small gas turbines to power a building or a campus. Since you're not losing electrons in transmission, it's highly efficient—and so it accounted for a third of all new U.S. generation in 2008, far outstripping nuclear or coal.[95] Windmills, too, can be in plain view, provided we can get past our reluctance to look at them, which has blocked their expansion in many of the most populated areas of the country. If we can see that spinning blade as something beautiful—as the breeze made visible, as the possibility for an energy future that might actually work—we may be able to make rapid progress. As with our food system, that progress would come faster if the government stopped subsidizing the fossil fuel industry and instead enacted policies like "feed-in tariffs" that force utilities to buy the juice from people's rooftops at a decent price. That's what Germany did, and as a result, despite its dreary and Wagnerian northern climate (Berlin is well north of Calgary), it boasts 1.3 million photovoltaic panels, more than any other country on earth.[96]

Solar panels are appearing not just in Frankfurt, or Denmark, or Portland, or Marin. In unlikely places. I spent a day not long ago in the Chinese city of Rizhao, a "newly emerging city" of 650,000, or roughly the size of Boston. It's clearly planning to get bigger; you come into town on an eight-lane boulevard plenty big for the six or eight cars meandering back and forth. A billboard at the municipal limits urges residents to "build a civilized city and be a civilized citizen." Part of being civilized is the chance to get clean—to take a shower now and then. To heat the water, you can build a coal-fired power plant and run an electric line to a hot-water heater in every basement. Or you could put solar hot-water heaters on each roof. In Rizhao, beginning in the 1990s, a few local entrepreneurs started selling the second option, and since Rizhao is a sunny city, it caught on. By now, virtually all the housing in the city heats water with the sun. I stood on the roof of one of its highest buildings, and on top of every apartment block I could see the distinctive long glass tubes. "It's over 95 percent," said Hu Yaibo, the city's general engineer, who was showing me around. "Some people say 99 percent, but I'm shy to say that." We ate lunch at a local hotel and then clambered up on the roof with the panting general manager. "See, this is a cloudy day and we're still making thirty-eight-degree Celsius water," he said. "On a sunny day, it's ninety degrees." China leads the world in installed renewable capacity; some of it is in big wind farms, but most of it can be found on top of buildings like these all over the country.

You can see the same kind of creativity across the developing world. I've stood in many tiny farm huts where the wife showed me, with considerable pride, the cooking flame provided by a biogas digester buried in the yard—pretty much a concrete tank, where you shovel the manure from a water buffalo or a cow or two. As that decomposes, it gives off enough gas for a rice cooker and to heat the water for a shower. It can change lives. Not far from the Great Wall, the agronomist Jules Pretty found a village-scale farm community that used not just the biogas digesters but also a small-scale gasification plant that burned corn husks to produce more methane. Instead of five hundred bushels a day of corn husks in the old inefficient stoves, this small machine required only twenty bushels. "Before women had to rush back from the fields to collect wood or husks," said the village leader. "And if it had been raining the whole house would be full of smoke. Now it is so clean and easy."[97]

You need to make use of what's at hand. I live on the boundary between forest and farm. I've already described how my neighbors have fanned out to spread new thermostats and lightbulbs, and how we've come together for wood-cutting bees to make sure everyone in town gets through winter plenty warm. But the most exciting project I've gotten to watch took shape at the college down the hill, Middlebury, which has emerged in recent decades as one of the nation's elite liberal arts institutions. "Liberal arts" is often one way of saying "not very practical," but Middlebury has the oldest environmental studies department in the nation, and a few years ago one environmental economics class undertook a carbon inventory of the college. They assembled a briefing book with 108 ideas for reducing emissions. Most were clever but small; soon the soda machines had motion sensors so they switched off their lights unless someone was standing nearby, and soon the golf carts on the college course sported electric motors. But not surprisingly, the analysis showed that four-fifths of the college's carbon emissions came from heating and cooling the dorms and dining halls and classrooms, on a campus that had sprawled considerably amid the boom of the 1990s. Middlebury was burning 2 million gallons of No. 6 fuel oil every year, and hence sending $4 million or $5 million off to the Persian Gulf or the Alberta tar fields. And so the college's staff looked around and noticed that Vermont, while short of hydrocarbons, was long in trees. In addition, a local company was making the highest-tech gasifying equipment in the world. So the trustees swallowed hard and spent the money for a $12 million wood-fired boiler that could handle half the college's load. It went online early in 2009.

When the switch was flipped, the college cut its fossil fuel use by 40 percent. That's the kind of number we need to start seeing if we're going to do anything about global warming. Just as important, it shortened its supply lines; all the wood chips come from within about seventy-five miles. But that's probably still not good enough; if many other institutions copied the system, we'd quickly deforest Vermont. So now agronomists have planted a fifteen-acre test plot of fast-growing willows down on the valley floor, on land that once pastured cows but was abandoned as the price of milk declined. Another environmental economics class is down there in the woods this semester. It looks like the trees are growing as fast as hoped, and that 2,400 acres should provide enough of a forest to heat and cool the college pretty much forever.

Is this the kind of thing that only rich liberal arts colleges might consider? Not really. When administrators were designing the plant, they drove forty miles north to Burlington, the state's biggest city, for a tour of the McNeil Generating Station, which has been burning wood to generate most of the city's power for decades. (It towers rustily above the Intervale farms, the 120-acre plot that supplies 8 percent of the city's produce.) In a given year, that power plant spends $15 million for local wood, providing lots of jobs for foresters; and because that money stays in the state, it creates, by one estimate, another 161 jobs.[98] (On the downside, that's $15 million less for some sheikh or strongman, doubtless depressing the private jet market.) The average Burlington customer paid less for electricity than his counterpart in the rest of New England, just as Middlebury should see its investment returned within five years.

Again—no silver bullets. Everyone can't pile into wood power, not even in heavily forested New England (where there are already plans on the books for hugely oversized biomass plants). As David Brynn, the director of Vermont Family Forests, pointed out, "Our forests produce just under four cords of wood per person in the state every year in terms of new growth," and about half that is already being harvested. "We are not awash in wood."[99] We'll need those new stands of willows—and we'll need the rapidly expanding "cowpower" initiative that takes methane from manure piles, not to mention the several thousand small hydro sites in the state. All this change would get much easier if the federal government favored small players, not huge corporations—try filling out the paperwork for your tiny run-of-the-river hydro plant. But it will also get much easier as the examples start to multiply, so that we begin to shake off our idea that energy is something that comes from a distance. That new wood-fired power plant at Middlebury was built with a forty-foot-tall glass window that lets everyone see the conveyor belts and boilers inside. If you come to tour the college, you'll see the new library but you'll also be shown the new boiler. And perhaps you'll reflect that they're alike in certain ways:

both based on wood fiber and both capable of giving off heat and light and sparks.

So now we're (theoretically anyway) well fed and warm. We can turn, then, to what may be the hardest part for most of us moderns to imagine about the future I'm describing, the greatest worry of all. If we're staying home, tending the garden, working with our neighbors, won't life be a tad . . . dull?

I know what I'm supposed to say. That if we renew our link to nature—to the cycle of its seasons—we'll be happier and healthier. That we were built for community—that it's been a great mistake to become so self-absorbed, and that we'll be relieved to depend once more on those around us. That living rooted to particular places is much sweeter than dwelling in the generic SoCal nowhere of the television. I even believe most of this; I've spent much of my life in tight communities and found great pleasure working hard alongside my neighbors or the other folks in my church. We are social animals, not far descended from the troops of great apes who spend all day grooming each other. I get disoriented if I leave home for very long. I know that in daily routine one can find deep satisfaction. I've read Thoreau. I've read Laura Ingalls Wilder. I grew up watching *The Waltons*.

On the other hand, *get real*. Most of us raised in the last fifty years are novelty junkies. A slower life, one not fixed on growth and change and speed but on a more plodding stability and security, will take a few generations to feel entirely comfortable. Our genes may yearn for it, but we were raised differently, and the transition, were it too abrupt, would be tough. There are Sundays when I just can't imagine another round of the same conversation I've had across the back of the church pew for the last decade. If that was all we had to look forward to, we'd feel shut in. We'd feel overwhelmed by the press of our neighbors. We'd be bored out of our skulls and simultaneously desperate for some time alone in our own heads.

Which is why we're lucky. The Internet may be precisely the tool we need; it's as if it came along just in time, a deus ex machina to make our next evolution bearable.

I was born in 1960, and in the years since then amazingly little has changed technologically. *Color* television, sure, but we had cars and antibiotics and airplanes and rocketships and telephones and washing machines when I came into the world. After the frenetic changes that marked the first half of the twentieth century, technology stood still. Except for the advent of the personal computer, and then the Internet, and then Google, which really did transform our daily lives. That is, one of the central tasks of my day—answering e-mail—didn't even exist twenty years ago; now the rhythm of my life changes appreciably on the weekend because the deluge of incoming messages slows to a manageable drip. Obviously, it's a mixed blessing. But it may be the exact blessing the moment demands.

A little math here. You have to plug in a computer; it requires energy. But a sip: a Harvard study early in 2009 found that "it takes on average about 20 milligrams of CO_2 per second to visit a web site."[100] About 1 percent of the world's electric supply goes to operating the data network.[101] The average Google search requires a few thousandths of a second of server time; you can make a thousand searches with the same amount of fossil fuel it takes to drive a car six-tenths of a mile. Which is important, because often those searches serve the same purpose as taking the car out for a spin; you're engaging with the world, finding something new. Venturing forth.

You could make a purely functional argument for the environmental value of the Internet, of course. If you have a computer, you can set up, say, a ride-sharing system that lets people coordinate their commutes or pick up a stranger on the way to the market. Or you can log on to the Freecycle network and find a way not to buy something. The Internet can take waste—that empty seat next to the driver, that old Ping-Pong table—and convert it into something useful.

But I'm thinking less tangibly. It's not so much the ride to town; it's the ride somewhere else entirely, into one of the millions of odd destinations that the Net provides. In its early days, people feared that the Web would simply turn into another television, dominated by a few big players. The media reformers Robert McChesney and John Nichols warned that "the internet was going to enhance concentration among media firms, as well as their overall size," that it would be "nearly impossible for anyone to start a commercially viable web site unless they are owned by or affiliated with an existing media giant."[102] Partly due to their own crusading work for "Net neutrality," that prediction proved wrong. Monoliths arise suddenly (Facebook, MySpace), and sometimes Rupert Murdoch buys them and sometimes they fade away, and meanwhile people just keep figuring out more ways to use a medium much more pliable than anything that came before. For one thing, it's cheap; that is, for very little money you can put up a Web site that looks worth reading. And it's oddly meritocratic—good stuff spreads quickly.

Mostly, though, it's *decentralized*. And that's why I like it. Just as food comes from the Midwest and oil from the Middle East, "content" used to come from New York and Hollywood. (And from a very small pool of people. Because I was president of the *Harvard Crimson* in college, there are days when all the bylines on the front page of the *New York Times* are people I drank beer with late at night in Cambridge barrooms.) All of a sudden it's possible to have the cultural equivalent of farmers' markets—content, ideas, craziness emerging from any place and every place. YouTube is a bazaar, Huffington Post is a souk. Some of it turns your stomach, and some is sublime.

That decentralization will be crucial, because all of a sudden we *need* vast amounts of information, very little of which can actually come from New York or Los Angeles. For instance, far more people are going to need to grow food. We're used to thinking of farmers as not quite bright, but in fact each one makes more decisions in a day than a whole platoon of investment bankers—and if they get many of them wrong, not much grows. That's why it's troubling that the chain of transmission for knowledge about how to farm is almost broken. In 2002, the average age of an American farmer was fifty-four, and steadily rising. Farmers' children have taken up other occupations. The big land-grant colleges have ag schools, but most of these might as well be subsidiaries of Cargill and Monsanto; they teach corporate-scale farming. So: it's a very good thing that when you google "how to compost video" you get about 1.7 million responses. There are thousands of films, many of them excellent. (Many of them dull, also, but earnest.) Myself, I like the one from kitchengardeners.org, with its companion presentation, "Our Buddy Bacteria." In the comments section, someone has written "Very good delivery of information. I shall act on it," which is precisely the response any of us communicators would yearn for.

For years, people continued to hope that television would somehow evolve into a teaching mechanism, but of course it never did; even if you made a program about how to compost, you'd reach only those people who happened to be watching at that particular time. The cost of making TV programs is so high you have to aim for the broadest possible audience, which at any given moment is less interested in compost than in, say, watching Ozzy Osbourne make a spectacle of himself. And if you try to make it on the cheap—community access TV!—it looks so terrible that no one can bear to watch. But a $199 camera makes a perfectly accessible YouTube video, which you can watch over and over again till you've got compost-making down pat. You can take your laptop out and perch it on the edge of your compost barrel if you need to. If you have a question, there are a thousand people you can quickly figure out how to ask. You can't learn *everything* via the Net. Really becoming a farmer, say, is much easier when you're surrounded by others doing the same thing, who can lend you tools and teach you how to use them. But, lacking a mother or an uncle who can show you how to do something, howto.com is a reasonable second choice.

The *Limits to Growth* authors believed that as ecosystems and economies began to decline, "service provision" would automatically start to fail as well. (They meant, say, the spread of information about birth control.) As I've said, the cost of treating dengue is probably already crowding out family planning clinics. But maybe this is one place we can be a little more optimistic; the spread of the Net seems to offer the chance for teaching literacy and even reaching the illiterate. Jules Pretty reports that Indonesian rice farmers have developed networks of tens of thousands that trade information on CDs; I've been in remote Asian villages where peasant farmers knew more about GATT and its effects than the average reader of the *Washington Post.* If you have a cell phone, you now know for yourself what the market price for your crop is; no need to rely on a middleman. It all helps.

And if you want to watch Ozzy Osbourne after a hard day out at the compost heap or the rice paddy? Well, that's the other thing about the Internet, of course. By now, essentially every minute of the twentieth century's electronic history is archived and available. You can stream tens of thousands of movies. Every vaguely musical sound emitted anywhere on the planet can be downloaded in a manner of seconds; every sitcom and dramedy and infomercial, every Japanese game show and Manchester United victory, every article in the history of the *New York Times*, every street on the planet cataloged in GoogleEarth. Every remnant of the age we've just come through—it's an inexhaustible mother lode of material. Just the stuff, perhaps, to slowly wean us away from the world we've known, to make sure we don't go cold turkey.

But also to wean us into the next world. Here's one of the oddest tricks the Net can teach: how to be a neighbor. When Michael Wood-Lewis and his wife, Valerie, moved to the south end of Burlington in 2000, he recalled, "We'd landed in what we thought was our dream neighborhood. It was walkable, near the lake, full of trees. But we were having trouble getting to know the neighbors. One night, my wife and I were sitting around the dinner table talking about it. It hit us that in the Midwest, where we were from, people brought cookies to their neighbors. We'd been here a year—where were our cookies?"[103]

Hence plan 1. They baked up a batch of Toll House specials and delivered them to the neighbors. "We used china plates, because I figured that way they'd have to return them and we'd get another conversation," Wood-Lewis remembered. "We never did get them back. I was kind of dumbfounded. But I don't think it was because people were rude. I think it's because people are living in a different culture than they were fifty years ago."

A culture busier and more distracted than before—busy enough that even in Vermont something had changed. So Wood-Lewis cooked up plan 2, which just may turn out to be one of the most innovative (and deceptively obvious) uses of the

Internet so far. In his hands, the Net has become a way to meet people not half a world away but half a block.

"I invested fifteen dollars at the copy shop and printed up four hundred flyers and put one on every door in our neighborhood. It pretty much just said, 'Share messages about lost cats and block parties.'" So was born the Five Sisters Neighborhood Forum, which he ran as a volunteer effort for six years. "It took about five minutes a day, and I was already on the computer anyway." Every evening he'd compile the five or six messages that had arrived at his in-box during the day, and send them out in a single e-mail bulletin—that was it. Someone would write in: "Neighbors, fyi: late last night I observed a large possum ambling across my front yard. Not as bad as a skunk, but I understand that possums can damage gardens and dig up lawns." Twenty-four hours later, another neighbor would have responded: "They have very soft feet that are not good for digging, and are not likely to cause lawn damage—and they are very clean animals and spend much of their rest time grooming themselves." Meanwhile, someone else has pruned their apple trees and wants to share the news that they have kindling piled up on the back porch free for the taking. Down the street someone's car has been broken into; the only thing taken was a gym bag filled with "my shoes, some sweaty clothes, and a couple of issues of the *New Yorker.* If anyone finds it dumped in their shrubbery, let me know."

Forget the World Wide Web—this one barely stretches four blocks. And no video, no rating systems, no celebrities, no hyperlinks. Just the daily rhythm of neighborhood life. "It grew steadily, from 10 or 20 percent of the neighborhood to the point where by 2006 we had 90 percent of the neighborhood signed up," said Wood-Lewis. That's when *Cottage Living* magazine included the area in its list of the ten best neighborhoods in the country. "And the reporter called me and he said, everywhere else in the country people would have dozens of different reasons why their place worked. But here, almost everyone put the e-mail thing on the list. That's what gave me the confidence."

The confidence to quit his job and start offering the service across all of Chittenden County, Vermont's largest. Within two years, FrontPorchForum.com reached thirteen thousand households, participating in more than a hundred neighborhood forums, some of them in inner-city neighborhoods where the main topics were how to fight graffiti and drive away drug dealers; some in rural towns where you get messages like: "We have four Indian Runner drakes who we expected to be females and lay beautiful round eggs. Instead we have these guys who really need some girls!!!"

This sounds like the stuff you'd see in the letters-to-the-editor column or on the bulletin board at the supermarket—and it is. But now it comes in an easy-to-use daily update that somehow breaks down barriers. "My sense was that this skill of neighborliness had eroded," said Wood-Lewis, citing data like the Harvard professor Robert Putnam's famous book *Bowling Alone*. "If you could increase social capital in a neighborhood—that is, your network of who you know and how well you know them—then your involvement increases. If you're among strangers, you're not going to volunteer for the Girl Scouts."

Sound theoretical? Not long after he'd launched his first forum, one of Wood-Lewis's neighbors was moving from an apartment to a house across the street. "They figured they could do it by themselves, but at the last minute decided they had a couple of big items they'd need some help with. So they put a note on the forum saying 'come Sunday at two'—and thirty-six people showed up. People didn't just move the chest of drawers and the bed; they organized into teams and boxed up the entire contents of the house, moved it across the street, and unpacked it, all in ninety minutes. I mean, someone pulled the picture hooks out of the wall in the old place and spackled over the holes. All the cardboard boxes were broken down and ready for recycling."

Front Porch Forum may already be the most important source of information for many Vermonters, who have watched their newspapers lay off reporters and shrink coverage. "One afternoon last year the state closed our main bridge as unsafe," recalled Erik Filkorn. "As a member of the town government I sent an extra to Michael Wood-Lewis, and he got the word right out. I think more people got the news that they'd have to change their morning commutes from him than from the traditional media." But it only works in emergencies because people use it every day; the steady stream of lost cats and people looking for summer jobs for their teenagers creates the community that people then rely on at more crucial moments. This same phenomenon is under way across the United States, as our declining economy has led to what sociologists call "an uptick of neighboring." At the University of Pennsylvania, Keith Hampton runs a Web site for community groups with over fifty thousand members, and the volume of messages grew 25 percent between 2008 and 2009. "I don't think people will create silos and hide in houses to shield themselves from hard times," he said. "They're going to look for people to help solve these problems. Those tend to be your neighbors."[104]

"There was a mother near us, with a teenage daughter who was having a birthday," Wood-Lewis recalled. "The girl wanted to go canoeing with her friends for her birthday, but when her mother checked out the price of renting canoes, it was too high. Her daughter said, 'I see lots of canoes in backyards around here,' but her mother was like, 'You can't just ask people you don't know for their boats.' Still, she put a one-line notice on the forum, saying they needed six canoes. Before the day was out, people were coming by. I mean, there were canoes just piling up in their front yard. She wrote me a note afterward: 'What a great feeling. What a great reminder of how to be a community. Why didn't I get to know these people ten years ago?' "

My point throughout this book has been that we'll need to change to cope with the new Eaarth we've created. We'll need, chief among all things, to get smaller and less centralized, to focus not on growth but on maintenance, on a controlled decline from the perilous heights to which we've climbed.

The saddest part of that notion, for me, is the chance that we'd need to leave behind not just the bad stuff but the good. I love my local community, but our national and global project has been about more than accumulation and expansion, more than cars and factories. It's also been about liberation—the slow but reasonably steady process of valuing more and more people. Valuing people from all around the world, men and women, people who think and love and look *different*. The universal solvent of open enterprise, of easy communication, of ever-increased mobility has broken down one wall after another, freed people in huge numbers from the life they would otherwise have been automatically assigned by their culture, their tradition, their gender. The process that made us anonymous to our neighbors carried real benefits, not just costs.

It's easy to see how, in a contracting world, that process could stall or even reverse—how aiming for a local economy and community solid enough to survive on this new planet might edge us back toward societies "traditional" in ways we don't want. We can no longer afford to ignore our neighbors; they'll be key to our survival. We're already seeing an increase in multigenerational families, as people can't afford nursing homes or bachelor pads; after a short hiatus, family ties may resume their historical role. Which is mostly good. It's nice to see Barack Obama's mother-in-law living in the White House; it feels right for the moment. But there are mother-in-law jokes for a reason, and people who live in small towns know that neighbors come with problems. They know that it's hard for new ideas to get in or old ones to get out. Such communities "generally demand conformity and punish intrusive eccentricity," writes Kirkpatrick Sale. "They are accustomed to fixed ways and patterns that do not allow much room for the excitement of novelty."[105] That's what *stifling* means.

Which is why, if I had my finger on the switch, I'd keep the juice flowing to the Internet even if I had to turn off everything else. We need cultures that work for survival—which means we need once more to pay attention to elders, to think hard about limits, to rein in our own excesses. But we also need cultures that work for everyone, so that women aren't made servants again in our culture, or condemned to languish forever as secondary citizens in other places. The Net is the one solvent we can still afford; jet travel can't be our salvation in an age of climate shock and dwindling oil, so the kind of trip you can take with the click of a mouse will have to substitute. It will need to be the window left ajar in our communities so new ideas can blow in and old prejudices blow out. Before, you had to choose between staying at home in the place you were born, with all its sensible strictures, and "going out in the world" to "make something of yourself." Our society—restless, mobile, wasteful, exciting, and on the brink—is the product of that dynamism. We can't afford to indulge those impulses anymore, but it doesn't mean we need to shut ourselves in.

Let me tell you a story about the last couple of years of my life, a story that I think makes this point in a small way. Twenty years ago I wrote the first book for a general audience about global warming, *The End of Nature,* and in the years since I've kept speaking and writing about climate. I'd been published in all the right places—the *New Yorker,* the *Atlantic, Harper's,* the *New York Review of Books, National Geographic*—but at some point I began to realize that this wasn't enough. It happened, I think, after that trip to Bangladesh where I came down with dengue and watched so many people dying. Something in me snapped. Nothing concrete had come of my work or anyone else's; Washington had done absolutely nothing to slow down climate change. I wanted to try and make something happen politically—but what? I'm a writer, living in the woods in Vermont. I called a few of my writer friends and suggested that we go up to Burlington and sit in on the steps of the federal building; we'd be arrested, and there'd be a small story in the paper. "Sounds good," they replied, as clueless as me. But one of them called the police in Burlington to ask what would happen, and the police said: nothing. Stay there as long as you want. We'll come visit.

Stymied, I simply started sending out e-mails to folks in my address book, saying, "Let's go for a walk, a kind of pilgrimage." With the help of a few friends who organized everything from food to first aid, it came together quickly. We stepped off three weeks later for a five-day trek through the mountains and valleys, sleeping in farm fields and holding evening programs in churches en route. When we got to Burlington, there were a thousand of us walking, which was enough to draw all our candidates for political office to our final rally. They didn't just talk; they all signed a big piece of cardboard we'd been carrying, pledging that they'd work to reduce carbon emissions 80 percent by midcentury. And not just the liberal Democrats. The woman who was running for Congress on the Republican ticket, and who almost won, had said in her campaign announcement some months earlier that she wasn't sure global warming was real, and that "more research was needed," the standard dodge. It turned out that the research wasn't in physics or chemistry—that the unanswered question was "How many people will walk across Vermont and ask me to change my mind?" And it turned out, empirically, that a thousand was enough, because she signed. The only downer was to read a story the next day in the paper that said those thousand people may have been the largest political rally solely against global warming that had ever happened in the United States. No wonder we were losing.

So we wondered if we could do something beyond funky Vermont. When I say "we," I mean myself and six seniors at Middlebury College. We had no money and no organization—no mailing lists, no fund-raising apparatus. But the six of them did have the kind of intuitive knowledge of the Internet that comes with being twenty-two years old. And so we sat around the college dining hall for long hours and talked about the possibilities. The people we consulted were helpful, but they were mostly veterans of an earlier era of social protest, and they kept talking about the need for a march on Washington. That seemed wrong, and not just because we lacked the chops to pull it off. For one thing, telling people to drive across the United States to protest global warming sounded weird. But the architecture of the United States had changed too, in a way that opened new possibilities. And so, in mid-January 2007, we again simply started sending e-mails to friends, asking them if they'd hold rallies and events of their own on April 14, about three months away, with that same goal: an 80 percent cut in emissions. And we asked them to forward the e-mail on to their friends. We had no idea how it would go. Our secret hope was that we'd organize a hundred of these demonstrations, which would have been a hundred more than there'd been before. But we didn't tell anyone, because it seemed grandiose and unlikely.

Instead, the idea took off. People started writing in from all over—they *got it.* They'd already changed their lightbulbs, and they wanted to do something more, but what? Global warming was the very definition of huge—how could anyone tackle it? But now they had an idea, a framework. Some came from big environmental groups (the Sierra Club, the National Wildlife Federation, the Natural Resources Defense Council, on and on). Far more were "just a housewife" or "just a student" or "just a pastor." It was as if we'd invited people to a potluck supper. They began figuring out how to make the idea of climate change real in their communities. We did our best to coordinate the gatherings, to organize materials, to do public relations work, to hold hands, all via e-mail. We'd told everyone that as soon as their rally was over they should upload pictures to our Web site, and we rented a hall in Washington, inviting all the politicians we could find to come watch the results. By early evening the photos were streaming in. It turned out that we'd "organized" 1,400 of these rallies in a single day, one of the largest days of grassroots environmental protest since the original Earth Day. We'd had our march on Washington, just in 1,400 different places.

But it wasn't just the number; it was the creativity. In Key West people held an underwater demonstration off the continent's only coral reefs, scuba divers and large fish cooperating to spread the news that global warming, unchecked, would destroy these glories. In Jacksonville they winched a yacht twenty feet into the air; when the Greenland icepack slides into the ocean, speakers pointed out, that is where the ocean will be. In lower Manhattan, thousands of people in blue shirts joined arms to form a human tideline and show where the ocean would rise across the country's most expensive real estate. People climbed glaciated peaks (which won't be glaciated much longer) and rallied in farmers' fields, in evangelical churches, on college campuses, in old folks' homes. It was unbearably moving to watch the pictures come in—this was distributed political action, the way that a farmers' market is distributed food production or a solar panel is distributed power. But because of the connecting power of the Web, it added up to more than the sum of its parts.

In fact, within a week, both Hillary Clinton and Barack Obama explicitly changed their energy and environment platforms, making 80 percent cuts in carbon emissions the centerpiece of their commitments. We hadn't done it all ourselves; Al Gore had laid the groundwork for real shifts in public attitude. But with $100,000 or so, and an almost endless amount of e-mail, we'd played a part. We felt good about ourselves—the technical term, I think, is *smug*.

Until about six weeks later, when the Arctic began to melt, and it became clear that almost everyone had underestimated the speed and size of the global warming now under way. These changes—the new Eaarth that I described in chapter 1—led, among other things, to James Hansen's NASA report setting 350 parts per million carbon dioxide as the maximum atmospheric concentration compatible with maintaining the planet "on which civilization developed and to which life . . . is adapted." And that, in turn, led to our small group (the young people to whom this book is dedicated) deciding to see if we could make our same tactic—distributed political action—work on a global basis.

Now the globe is a big place, and people insist on speaking different languages; there's a reason that the Golden Arches and the Coke bottle about exhaust the list of global symbols. But we decided to take those three digits—350—and see what we could do. With the world about to conclude a new set of treaties on carbon, we knew that we needed a strong target to force real action. So the goal was simply to drive those numbers into hearts and minds around the world. We found other college kids who could translate the message into a dozen languages, and we started writing e-mails. And, just as before, people began to respond. In fact, the word spread even more quickly this time; suddenly there were e-mails appearing from, say, a farmer in a village in Cameroon, who had heard about the campaign from a text message over his mobile phone. (You may think the Internet is a Western thing, but you have to go pretty far back of beyond to find a village without a cell phone. And people have learned to use those mobiles to spread all kinds of information—our ninety-second animated wordless video looked great on cell phone screens.) Anyway, he and his neighbors had understood what we were getting at; they planted 350 trees on the edge of the village, hung a small sign, and took a picture.

This gesture nearly made me weep. People in Cameroon have done nothing to cause global warming; they will be hammered by its effects; but now they feel like they have some small way of fighting back. Fifteen hundred Buddhist monks and nuns formed a huge human "350" against the backdrop of the Himalayas in Ladakh; great athletes formed a "team 350," with NFL stars and champion skiers and Tour de France racers; a picture arrived of a climber atop the highest mountain in Antarctica with a huge

"350" banner. Soon we had "offices" all over the globe: Judit ran the show in Hungary, Wa-el in Beirut, Samantha in Johannesburg, Ely in the Congo, Govind in Delhi, Abe in Malaysia, Aaron in New Zealand, Paolo in Quito. And in the summer of 2009 Rajendra Pachauri, the United Nations' top climate scientist, who had accepted the Nobel Prize alongside Al Gore for his work in leading the Intergovernmental Panel on Climate Change, endorsed our goal. "What is happening, and what is likely to happen," he said, "convinces me that the world must be really ambitious and very determined at moving toward a 350 target."[106]

We set our sights on a huge day of global action in October 2009, with actions everywhere from the highest mountains to underwater on the Great Barrier Reef. There were flotillas of canoes and lines of bikes; people in great human chains along beaches; hundreds of churches pealing their bells 350 times. It was, said CNN, "the most widespread day of political action in the planet's history." No one needed to leave his city or town; no one needed to march on Washington or Paris or Cairo. People simply showed their heart and soul and creativity where they were and then used this new tool, the Internet, to become larger than the sum of the parts. We didn't need to own a TV network; for a day, the number hung in the air. Almost literally—we had satellites taking pictures of the biggest demonstrations. If someone was looking on from a different galaxy, they saw, for an afternoon, that humans had figured out something about limits. Something about the planet where they lived.

It's true that by some measures we started too late, that the planet has changed and that it will change more. The president of the Maldives held an underwater cabinet meeting to pass a 350 resolution, but the president of the United States didn't convert overnight into a passionate supporter, for all the reasons I've described. The momentum of the heating, and the momentum of the economy that powers it, can't be turned off quickly enough to prevent hideous damage. But we will keep fighting, in the hope that we can limit that damage. And in the process, with many others fighting similar battles, we'll help build the architecture for the world that comes next, the dispersed and localized societies that can survive the damage we can no longer prevent. Eaarth represents the deepest of human failures. But we still must live on the world we've created—lightly, carefully, gracefully.

• AFTERWORD •

The months after the initial publication of *Eaarth* saw some of the most intense environmental trauma the planet has ever witnessed, events that exemplified the forces I have described in the book.

For Americans, the British Petroleum (BP) oil spill in the Gulf of Mexico, which began on April 20, 2010, may have provided the most powerful images—there was, after all, an underwater camera showing the leak up close. (Leak? This was not a leak—it was a stab wound that BP inflicted on the ocean floor, a literal hole in the bottom of the sea. If you ever had any doubts about peak oil, all it took was one view of the extreme places and pressures the oil companies now had to endure to find even marginal amounts of crude. The well that BP was drilling would have supplied only about four days' worth of America's oil consumption.) The pictures reminded us of the thing we've been trying to forget since Rachel Carson published *Silent Spring* nearly fifty years before: "Progress" and "growth" come with a dark side, in this case an easy-to-see dark black side. Just a couple of weeks before the spill, President Barack Obama had reopened much of the coastline to oil drilling, arguing, "It turns out, by the way, that oil rigs today generally don't cause spills. They are technologically very advanced." By midsummer, a chagrined president was reduced to telling the nation that he'd only lifted the moratorium "under the assurance that it would be absolutely safe."

But that's the point—there's nothing absolutely safe anymore, not when we're pushing past every limit. There's not even anything relatively safe; we're overloading every system around us. If it's not too big to fail, it's too deep to fail, or too complicated to fail. And it's failing.

As it turns out, however, the BP spill was not the most dangerous thing that happened in the months after this book was first published. In fact, in the spring and summer of 2010, the list of startling events in the natural world included:

- Nineteen nations setting new all-time high temperature records, which in itself is a record. Some of those records were for entire regions—Burma set the new mark for Southeast Asia at 118 degrees, and Pakistan the new zenith for all of Asia at 129 degrees.
- Scientists reported that the earth had just come through the warmest six months, the warmest year, and the warmest decade for which we have records; it appears 2010 will be the warmest calender year on record.
- The most protracted and extreme heat wave in a thousand years of Russian history (it had never before topped 100 degrees in Moscow) led to a siege of peat fires that shrouded the capital in ghostly, deadly smoke. The same heat also cut Russia's grain harvest so sharply that the Kremlin ordered an end to all grain exports to the rest of the world, which in turn drove up world grain prices sharply.
- Since warm air holds more water vapor than cold air (as explained in chapter 1), scientists were not surprised to see steady increases in flooding. Still, the spring and summer of

2010 were off the charts. We saw "thousand-year storms" across the globe, including in American locales like Nashville, the mountains of Arkansas, and Oklahoma City, all with deadly results. But this was nothing compared with Pakistan, where a flooded Indus River put 13 million people on the move, and destroyed huge swaths of the country's infrastructure.

- Meanwhile, in the far north, the Petermann Glacier on Greenland calved an iceberg four times the size of Manhattan.
- And the most ominous news of all might have come from the pages of the eminent scientific journal *Nature,* which published an enormous study of the productivity of the earth's seas. Warming waters had put a kind of cap on the ocean, reducing the upwelling of nutrient-rich cold water from below. As a result, the study found, the volume of phytoplankton had fallen by half over the last sixty years. Since phytoplankton is the world's largest source of organic matter, this was unwelcome news.

Indeed, all of these observations were unwelcome, if at some level expected. They were further, deeper signs of earth transforming itself into Eaarth. And they had reached the level where few who lived through the events wanted to deny their meaning. Here was the president of Russia, Dimitri Medvedev, after watching the fires that shut down Moscow for weeks: "Everyone is talking about climate change now. Unfortunately, what is happening now in our central regions is evidence of this global climate change, because we have never in our history faced such weather conditions in the past." (This from the president of a country whose economy totally depends on the endless production of oil and gas.)

And here was *The New York Times,* which had spent years piously explaining that there were two sides to the question of warming. In mid-August 2010, above the fold on a Sunday paper, the *Times* ran three huge photos of flood, melt, and fire, and beneath them a story that declared: "These far-flung disasters are reviving the question of whether global warming is causing more weather extremes. The collective answer of the scientific community can be boiled down to a single word: probably." Okay, probably is still a weasel word—but for the *Times,* a breakthrough. "The warming has moved in fits and starts, and the cumulative increase may sound modest," the paper reported. "But it is an average over the entire planet, representing an immense amount of added heat, and is only the beginning of a trend that most experts believe will worsen substantially."

There is no satisfaction at all in saying I told you so. I've been saying it for two decades, ever since the publication of *The End of Nature,* and it's never been sweet in the slightest. I'd give a lot to have been wrong instead.

But if there's one development that chafes above all others, it's political: the decision by the U.S. Congress in the summer of 2010 to punt, spectacularly, on doing anything about climate change. During the Bush years, of course, inaction had been a given. But with the advent of Democratic majorities in Congress and then the election of Barack Obama, some hope emerged that Washington might decide to act. That action would never have been dramatic or decisive; in June 2009 the House passed a weak bill that would scarcely have cut emissions in the next few crucial years. But at least it was *something,* a token effort that might have boosted the world's morale enough to help put international climate negotiations back on track, even after the debacle at Copenhagen six months later. When the legislation reached the Senate, however, it stalled for more than a year. Big coal and big oil didn't care for it, and so their squads of lobbyists went to work. Senator John Kerry of Massachusetts, who was leading the charge on the legislation, didn't so much charge as retreat, again and again and again. Here's how he put it on the eve of the final battle: "We believe we have compromised significantly, and we're prepared to compromise further." With leadership like that, what could possibly go wrong?

Meanwhile, the White House did nothing that might have added to the pressure for change. Instead of using the horrible BP spill as a reason to act, President Obama failed to draw the obvious connection: that fossil fuel is dirty stuff, whether it spills into the Gulf from a broken well or spills into the atmosphere from the tailpipes of our cars. With no help from the administration, the outcome was such a given that the Senate decided not even to vote—the members of the "world's greatest deliberative body" simply walked away. The best guess of various observers was that, after the GOP sweep in the midterm elections, we may have to wait until 2013 to see another legislative opening.

As readers of *Eaarth* know, I think it's unlikely that bills of the scale proposed in Washington, or agreements of the magnitude considered in Copenhagen, will make any substantive difference in the outcome. Our leaders have failed to come to terms with the actual size of the problem: that unless we commit ourselves to a furious push to get back to 350 parts per million, the damage will be overwhelming. (The scariest thing about the scary summer of 2010 was that it happened with only one degree of warming, globally averaged; we face five or six degrees this century if we don't take crisis action to get off fossil fuel.) Hence, in some sense, the failure of these various legislative efforts is disgusting but not decisive.

In certain ways, in fact, it clears the air. For years, the effort to build a movement to do something about climate change in the

United States has been hampered by the presence of these weak bills. It was hard to rally people to a banner when that banner hung so limply. Now that there is no real chance of tough action in the next year or two, a real opportunity exists to build a powerful, angry movement, in the United States and around the world—a movement capable of pushing for real change on a scale that matters. That's what organizations like 350.org are trying to do, exploring strategies that range from planet-scale art projects to concerted civil disobedience.

At the same time, since we're not going to forestall some really disastrous climate change, the need to make communities more resilient continues apace. And sometimes these two thrusts can be combined. On October 10, 2010, 350.org coordinated 7,400 different actions in 188 countries into a Global Work Party. In far-flung places, people put up solar panels, dug community gardens, laid out bike paths—all the kinds of things that will help make those places more likely to endure in a warmer world. But they also used the occasion to send a strong political message to their leaders. At day's end, they put down the shovels, picked up cellphones, and left the same message: "We're getting to work, what about you?"

These are the two strands we must simultaneously undertake. We've got to harden our communities so they can withstand the couple of degrees of global warming that are now inescapable. (And, as the summer of 2010 showed, that's no easy task.) At the same time, we've got to cooperate internationally to force legislative change that will hold those increases below the four or five degrees that would make a difficult century an impossible one. Much will depend on how effective that movement-building turns out to be.

• • •

In the end, the BP spill that dominated the headlines for much of the summer of 2010 turned out to be the less important sign of environmental crisis, and not only because its effect was smaller than, say, the Pakistani flooding. It's because the BP spill was an *accident*. It was a one-off crisis: The proper response was to ban deepwater drilling till we know how to do it, to make sure all wells are as safe as possible, and to compensate fully everyone damaged by BP's greed. In that sense, it fit in with our legacy idea of what constitutes pollution: something going wrong.

But the greatest danger we face, climate change, is no accident. It's what happens when everything goes the way it's supposed to go. It's not a function of bad technology, it's a function of a bad business model: of the fact that Exxon Mobil and BP and Peabody Coal are allowed to use the atmosphere, free of charge, as an open sewer for the inevitable waste from their products. They'll fight to the end to defend that business model, for it produces greater profits than any industry has ever known. We won't match them dollar for dollar: To fight back, we need a different currency, our bodies and our spirit and our creativity. That's what a movement looks like; let's hope we can rally one in time to make a difference.

• NOTES •

PREFACE

1. U.S. Climate Change Science Program, "Weather and Climate Extremes in a Changing Climate; Synthesis and Assessment Product 3.3," June 2008, p. 4.

2. "Deluge, Blizzard—These Are Wetter Times," *Foster's Daily Journal* (New Hampshire), January 26, 2009.

3. Richard Ingham, "Act Now on Floods, Drought, Says Forum," *Age* (Australia), March 18, 2009.

4. Pavel Y. Groisman, Richard W. Knight, and Thomas R. Karl, "Heavy Precipitation and High Streamflow in the Contiguous United States: Trends in the Twentieth Century," *Bulletin of the American Meteorological Society* 82, no. 2 (February 2001): 219.

5. "Suffering the Science: Climate Change, People, and Poverty," Oxfam Briefing Paper 130, July 6, 2009, p. 1.

1: A NEW WORLD

1. Andrew Revkin, "Puberty on the Scale of a Planet," *New York Times*, August 7, 2009.

2. Robert Poole, "For the Apollo Astronauts, A Small World," *Los Angeles Times*, July 19, 2009.

3. Rosslyn Beeby, "Warming Fuels Rise in Tropical Storms," *Canberra Times*, December 27, 2008.

4. "NASA Study Links Severe Storm Increases, Global Warming," *Pasadena Star News*, January 23, 2009.

5. Brian K. Sullivan, "California Fire Season Now Year Round in Era of Mega Blazes," Bloomberg.com, November 18, 2008.

6. Jonathan Leake, "Arctic Ice Melting Even in Winter," *Times* online, October 26, 2008.

7. "First Commercial Ship Sails through Northwest Passage," *Climate Progress*, November 30, 2008.

8. Lisa Jarvis, "Kindling for Climate Change," *Chemical and Engineering News*, August 17, 2009.

9. Mason Inman, "Arctic Ice in 'Death Spiral,'" *National Geographic News*, September 17, 2008.

10. Joseph Romm, "Loss Cause," Grist.org, December 19, 2008.

11. "The Curse of Carbon," *Economist*, December 31, 2008.

12. Steve Connor, "Expanding Tropics a Threat to Millions," *Independent*, December 3, 2007.

13. Rachel Kleinman, "No More Drought: It's a 'Permanent Dry,'" *Age*, September 7, 2007.

14. David Pallister, "Brushfires and Global Warming, Is There a Link?" *Guardian*, February 8, 2009.

15. David Pallister, "Australian Brushfire Toll Its Worst Ever," *Guardian*, February 8, 2009.

16. Patty Henetz, "Drought Deepens Strain on a Dwindling Colorado," *Salt Lake Tribune*, November 29, 2008.

17. Mike Stark, "Climate Change, Drought to Strain Colorado River," Associated Press, December 5, 2008.

18. Abraham Lustganta, "How West's Energy Boom Could Threaten Drinking Water," *San Diego Tribune*, December 21, 2008.

19. World Wildlife Fund, "Climate Change: Faster, Stronger, Sooner," October 20, 2008.

20. Tania Banigan, "Drought Threatens China Wheat Crop," *Guardian*, February 4, 2009.

21. Michael Klare, "The Second Shockwave," *Foreign Policy in Focus*, March 18, 2009.

22. Suzanne Goldenberg, "Climate Change Threatens Ganges, Niger and Other Mighty Rivers," *Guardian*, April 22, 2009.

23. Kyangjin Gompa, "Himalayan Villages on Global Warming Frontline," Agence France-Presse, December 26, 2008.

24. "On Thinner Ice: Melting Glaciers on the Roof of the World," Asia Society, http://www.asiasociety.org/onthinnerice.

25. Raju Gusain, "Climate Change Leads to Early Flowering," *India Today*, February 4, 2009.

26. Tim Rippel, "Slippery Slope," *Hemispheres*, April 2009, p. 63.

27. John Enders, "Bolivia's Chacaltaya Glacier Is Gone," *Miami Herald*, May 5, 2009.

28. Doug Struck, "On the Roof of Peru, Omens in the Ice," *Washington Post*, July 29, 2006.

29. Rick Jervis, "Data Show U.S. Riding Out Worst Storms on Record," *USA Today*, October 22, 2008.

30. NTS Asia-Secretariat, "Climate Refugees, A Crisis in the Making," October 2008.

31. Charles M. Blow, "Farewell, Fair Weather," *New York Times*, May 31, 2008.

32. Julian Siddle, "Marine Life Faces 'Acid Threat,'" BBC, November 30, 2008.

33. Robert Lee Hotz, "A Look into Future Oceans for Shellfish Reasons," *Wall Street Journal*, April 4, 2009.

34. "The Curse of Carbon," *Economist*, December 31, 2008.

35. Craig Welch, "Oysters in Deep Trouble," *Seattle Times*, June 14, 2009.

36. John Aglionby, "Scientists Fear for Seas at Climate Talks," FT .com, May 14, 2009.

37. Douglas Fischer, "The Ocean's Acid Test," DailyClimate.com, November 12, 2008.

38. "Transcript: John McCain's Foreign Policy Speech," *New York Times*, March 26, 2008.

39. Paul Roberts, *The End of Oil* (Boston: Houghton Mifflin Harcourt, 2004), p. 125.

40. Connor, "Expanding Tropics."

41. John Rockstrom et al., "A Safe Operating Space for Humanity," *Nature* 461 (September 27, 2009): 472–75.

42. A. Tripathi et al., "Coupling of CO_2 and Ice Sheet Stability," *Science*, October 8, 2009.

43. Zoological Society of London, "Coral Reefs Exposed to Imminent Destruction from Climate Change," news release, July 6, 2009.

44. William Branigin, "Obama Plans to Overhaul Environmental Policies," *Washington Post*, January 26, 2009.

45. Richard Harris, "Global Warming Is Irreversible, Study Says," *All Things Considered*, January 26, 2009.

46. David Adam, "Too Late," *Guardian*, December 9, 2008.

47. "U.S. CO_2 Emissions Are Falling," Reuters, August 11, 2009.

48. Jonathan Weisner, "Global Warming Goal Set," *Wall Street Journal*, July 3, 2009.

49. Steve Connor, "Exclusive: Methane Time Bomb," *Independent*, September 23, 2008.

50. Volker Mrasek, "A Storehouse of Greenhouse Gases Is Opening in Siberia," *Spiegel* online, April 17, 2008.

51. Melissa Block, "Scientist Measures an Overlooked Greenhouse Gas," *All Things Considered*, September 10, 2007.

52. Fred Pearce, "Arctic Meltdown Is a Threat to Humanity," *New Scientist*, March 25, 2009.

53. Joe Romm, "For Peat's Sake: A Point of No Return as Alarming as the Tundra Feedback," ClimateProgress.org, October 13, 2008.

54. "Sinking Feeling: Hot Year Damages Carbon Uptake by Plants," Agence France-Presse, September 17, 2008.

55. David Adam, "Sea Absorbing Less CO_2, Scientists Discover," *Guardian*, January 12, 2009.

56. Global Carbon Project 2008, "Carbon Budget and Trends 2007," September 26, 2008.

57. http://www.climatemediapartnership.org/spip.php?article709.

58. Lester Brown, "Could Food Shortages Bring Down Civilization," *Earth Policy News*, September 29, 2009.

59. Martin Mittelstaedt, "Unprecedented Heat Will Trigger Global Food Crisis," *Toronto Globe and Mail*, January 9, 2009.

60. Lewis Smith, "Billion People Face Famine by Mid-Century, Says Top US Scientist," *Times* (London), March 23, 2009.

61. Maggie Fox, "Climate Warming Means Food Shortage, Study Warns," Reuters, January 9, 2009.

62. "Death Bloom of Plankton a Warning," *San Francisco Chronicle*, November 21, 2008.

63. Ker Than, "¾ of Big Antarctic Penguin Colonies to Disappear?" *National Geographic News*, December 1, 2008.

64. Catherine Brahic, "Honey, Climate Change Is Shrinking the Species," *New Scientist*, September 11, 2008.

65. Alex Morales, "Sheep Shrinking Each Generation Amid Global Warming," Bloomberg.com, July 3, 2009.

66. Zoe Cormeir, "It's Attack of the Slime," *Toronto Globe and Mail*, December 27, 2008.

67. Silvia Aloisi, "Ocean Noise on Increase," Reuters, December 3, 2008.

68. Christine Dell'Amore, "Giant Mucus-like Sea Blobs on the Rise," *National Geographic News*, October 8, 2009.

69. Doyle Rice, "Global Warming May Be Twice as Bad as Previously Expected," *USA Today*, May 21, 2009.

70. Nate Hagens, interviewed by Marianne Lavelle, in "Beyond the Barrel," January 7, 2008, *US News and World Report* online.

71. Richard Heinberg, "George W. Bush and Peak Oil: Beyond Incompetence," *Energy Bulletin*, March 21, 2006.

72. Neil King Jr. and Spencer Swartz, "Oil Supplies Will Tighten, IEA Warns," *Wall Street Journal*, November 7, 2008.

73. Joe Romm, "Merrill: Non-OPEC Production Has Likely Peaked," ClimateProgress.org, February 2, 2009.

74. http://www.energybulletin.net/node/46556.

75. Travis Bradford, *Solar Revolution* (Cambridge, Mass.: MIT Press, 2006), p. 40.

76. Richard Heinberg, *The Oil Depletion Protocol* (Gabriola Island, British Columbia: New Society, 2006), p. 7.

77. Roberts, *The End of Oil*, p. 153.

78. Ibid., p. 28.

79. Kurt Cobb, "The Net Energy Cliff," *Energy Bulletin*, September 14, 2008.

80. Rob Hopkins, *The Transition Handbook* (White River Junction, Vt.: Chelsea Green, 2008), p. 51.

81. Phil Hart, theOilDrum.com, October 23, 2008.

82. Nate Silver, "The End of Car Culture," *Esquire*, May 14, 2009.

83. "Did the Oil Price Boom of 2008 Cause Crisis?" *Wall Street Journal*, WSJ.com, April 3, 2009.

84. George Soros, "The Perilous Price of Oil," *New York Review of Books*, September 25, 2008.

85. "Fears for Landmark Bridge," *Independent Television News*, August 31, 2008.

86. "Widespread Flooding Forces State of Emergency in Marshall Islands," Agence France-Presse, December 25, 2008.

87. http://healthvermont.gov/news/2008/060508lyme.aspx.

88. Alyssa Abkowitz, "Beating Back the Ocean Proves an Enduring Riddle," *Wall Street Journal*, September 12, 2008.

89. Natural Resources Defense Council, "The Consequences of Global Warming," September 21, 2007.

90. "New Report—Climate Change Threatens Ohio," environmentohio.org, December 17, 2008.

91. Tom Henry, "Climate Change Called Certain and Most Predictions Are Bad," *Toledo Blade*, October 13, 2008.

92. Ibid.

93. John Vidal, "Wetter and Wilder: The Signs of Warming Everywhere," *Guardian*, December 10, 2008.

94. Eliza Barclay, "Peru's Potato Farmers Adapt to Climate Change," *Miami Herald*, September 15, 2008.

95. Ben Simon, "Lifestyle Melts Away with Uganda Peak Snow Cap," Agence France-Presse, June 15, 2009.

96. Marc Lacey, "Meager Living of Haitians Is Wiped Out by Storms," *New York Times*, September 11, 2008.

97. Jack Healy, "A Luxury Cruise in Harm's Way," *New York Times*, December 3, 2008.

98. http://www.findingdulcinea.com/news/international/2008/November/Maldives-May-Relocate-Due-to-Global-Warming.html.

99. Subramian Sharma, "Kiribati Islanders Seek Land to Buy as Rising Seas Threaten," Bloomberg.com, February 9, 2009.

100. http://www.oceaniacruises.com/T_MainContentPage.aspx?PageUID=dc6fb51a-8819-465a-93b5-30aec64cde17.

101. Tom Phillips, "Brazil: Deforestation Rises Sharply as Farmers Push into Amazon," *Guardian*, September 1, 2008.

102. Rex Weyler, "Deep Green: Forests, Carbon Sink or Carbon Bomb?" Greenpeace.org, May 14, 2009.

103. Anna Armstrong, "Fiery Forecast," Nature Reports Climate Change, November 27, 2008.

104. Peter Bunyard, "Climate Change and the Amazon," in Herbert Girardet, *Surviving the Century* (London: Earthscan, 2007), p. 85.

105. Ibid., p. 91.

106. Peter Bunyard, "Gaia, Climate and the Amazon," July 15, 2005, http://www.indsp.org/SWPeterBunyard.php.

107. Tom Knudson, "Sierra Warming, Later Snow, Earlier Melt," *Sacramento Bee*, December 26, 2008.

108. Doug Bartholomew, "Experts Planning for a Flood of Noah's Ark Proportions," Dailybulletin.com, July 27, 2009.

109. Knudson, "Sierra Warming."

110. Matthew Daly, "House Approves Special Spending to Fight Wildfires," Associated Press, March 26, 2009.

111. Tom Knudson, "Sierra Nevada Climate Changes Feed Monster, Forest-Devouring Fires," *Sacramento Bee*, November 30, 2008.

112. Matt Walker, "Yosemite's Giant Trees Disappear," BBC News, May 30, 2009.

113. Mireya Navarro, "Environment Blamed in Western Tree Deaths," *New York Times*, January 23, 2009.

114. Ed Stoddard, "Forests Fall to Beetle Outbreak," Reuters, August 4, 2009.

115. Jim Robbins, "Bark Beetles Kill Millions of Acres of Trees in the West," *New York Times*, November 18, 2008.

116. Scott LaFee, "Pining Away," San Diego.com, December 21, 2008.

117. http://dl.klima2008.net/ccsl/ccf_report_oct_06.pdf.

118. Eric Newhouse, "Climate Change Affecting Mountains Most," *Great Falls Tribune*, February 9, 2009.

119. Stephen Speckman, "Bark Beetles Are Feasting on Utah Forests," *Desert News*, September 8, 2008.

120. W. A. Kurz, "Mountain Pine Beetle and Forest Carbon Feedback," *Nature* 452 (April 24, 2008): 987–90.

121. Howard Witt, "Canada's Forests, Once Huge Help on Greenhouse Gases, Now Contribute to Climate Change," *Chicago Tribune*, January 2, 2009.

122. Scott Streater, "Climate Change, Water Shortages Conspire to Create 21st Century Dust Bowl," *New York Times*, May 14, 2009.

123. Juliet Eilperin, "Dust Storms Escalate, Prompting Environmental Fears," *Washington Post*, April 23, 2009.

124. Streater, "Climate Change."

125. Eilperin, "Dust Storms Escalate."

126. "Climate Change Threatens Lebanon's Legendary Cedars," Agence France-Presse, February 5, 2009.

127. James Kantner, "Global Tourism and a Chilled Beach in Dubai," *New York Times*, January 2, 2009.

128. Peter Ker, "Water Plant to Guzzle Energy," *Age*, August 30, 2008.

129. Moslem Uddin Ahmed, "Water Everywhere, But Not a Drop to Drink," *New Nation* (Dhaka), October 27, 2008.

2: HIGH TIDE

1. Robert M. Collins, *More: The Politics of Economic Growth in Postwar America* (New York: Oxford University Press, 2000), p. 8.

2. Kevin G. Hall, "CBO: Obama's Budget Would Double National Debt," *San Jose Mercury News*, March 20, 2009.

3. Michael Scherer, "Will Deficits Force Obama to Sacrifice His Agenda," *Time*, March 23, 2009.

4. Transcript, White House press conference, March 24, 2009.

5. Kevin Hall, "Rosy Predictions Underlie Obama's Budget," *Bellingham Herald*, February 26, 2009.

6. Ban Ki-moon and Al Gore, "Green Growth Is Essential to Any Stimulus," *Financial Times*, February 16, 2009.

7. "Green Growth: Korea's New Strategy," *Korea Herald*, March 24, 2009.

8. Thomas L. Friedman, *Hot, Flat, and Crowded* (New York: Farrar, Straus and Giroux, 2008), p. 24.

9. Ibid., p. 25.

10. Thomas L. Friedman, "Time to Reboot America," *New York Times*, December 23, 2008.

11. http://74.125.47.132/search?q=cache:ymD6YNNNIswJ:www.un.org/partnerships/Docs/Press%2520release_Google_17Jan08.pdf.

12. Friedman, *Hot, Flat, and Crowded*, p. 231.

13. Ibid., p. 176.

14. "EIA Projects Wind at Five Percent of United States Electricity in 2012," DemocraticUnderground.com, May 26, 2009.

15. Vaclav Smil, "Moore's Curse and the Great Energy Delusion," *American*, November 19, 2008.

16. http://www.hybridcars.com/2009-hybrid-cars.

17. Jerome Guillet, "Official Energy Reports (2)," Eurotrib.com, November 12, 2008.

18. Daniel J. Weiss and Alexandra Kougentakis, "Big Oil Misers," Center for American Progress, March 31, 2009, http://www.americanprogress.org/issues/2009/03/big_oil_misers.html.

19. Smil, "Moore's Curse," p. 9.

20. Ibid., p. 10.

21. Paul Roberts, *The End of Oil* (Boston: Houghton Mifflin Harcourt, 2004), p. 132.

22. Ibid., p. 270.

23. "Forty-four Percent Say Global Warming Due to Planetary Trends," Rasmussenreports.com, January 19, 2009.

24. Jad Mouawad, "Green Is for Sissies," *New York Times*, November 16, 2008.

25. Marianne Lavelle, "The Climate Change Lobby Explosion," Center for Public Integrity, February 24, 2009.

26. Scott Malone, "Pickens' Downgraded Plans May Reflect Shift in Wind," Reuters, July 8, 2009.

27. Joe Romm, "Exclusive Analysis, Part 1: The Staggering Cost of New Nuclear Power," ClimateProgress.org, January 5, 2009.

28. Craig A. Severance, "Business Risks and Costs of New Nuclear Power," ClimateProgress.org, January 2, 2009.

29. Mariah Blake, "Bad Reactors," *Washington Monthly*, January–February 2009, p. 31.

30. Joe Romm, "Warning to Taxpayers, Investors, Part 2," ClimateProgress.org, January 7, 2009.

31. Romm, "Staggering Cost of New Nuclear Power."

32. Joe Romm, "How the World Can and Will Stabilize," ClimateProgress.org, March 26, 2009.

33. Brittany Schell, "Nukenomics No Longer Add Up—Expert," OneWorld.com, October 31, 2008.

34. Romm, "How the World Can and Will Stabilize."

35. Joann Loviglio, "Report: United States Bridges Falling Down," Associated Press, August 27, 2008.

36. Scott Hadly, "United States Infrastructure Shaky, Official Says," *Ventura County Star*, November 29, 2008.

37. "America's Crumbling Infrastructure Requires a Bold Look Ahead," *Economist*, July 3, 2008.

38. Bob Kinzel, "Road Repair Cost Projection Higher than Anticipated," Vermont Public Radio, September 18, 2007.

39. Larry Wheeler and Grant Smith, "Pipeline of Trouble," *USA Today*, August 27, 2008.

40. Mike DeSouza, "Cities Threatened by Global Warming," *Calgary Herald*, December 16, 2008.

41. NoLa.com, "Protecting Southeast Louisiana Will Be Extraordinarily Expensive," December 15, 2008.

42. Angela Ellis, "Future Storms Could Devastate Louisiana Coast," ABC News, August 28, 2008.

43. "One Million Bangladesh Cyclone Survivors Await Homes—Oxfam," Reuters, November 12, 2008.

44. Pacific Disaster Center, "World's Smallest Island Nation Faces Uncertain Future," *PDC Analysis* 1, no. 2 (January 2004): 3.

45. Rory Carroll, "We Are Going to Disappear One Day," *Guardian,* November 8, 2008.

46. "Cuba Readies for Another Hurricane," Reuters, November 8, 2008.

47. "Historic Center of Venice Flooded," *International Herald Tribune,* December 1, 2008.

48. Juan A. Lozano, "Texas Mulls Massive 'Ike Dike,'" Associated Press, July 15, 2009.

49. Alix Rijckaert, "Dutch Government Warned Against Rising Sea Levels," Agence France-Presse, September 3, 2008.

50. Liz Mitchell, "Are Rising Sea Levels a Bigger Threat Here than Hurricanes," *Island Packet*, October 31, 2008.

51. Tom Ramstack, "Warming Scenario Sees Flooded Airport," *Washington Times,* July 8, 2008.

52. Ron Friedman, "Israel Urged to Act Now," *Jerusalem Post,* July 6, 2007.

53. Cornelia Dean, "United States Infrastructure at Risk from Rising Seas," *International Herald Tribune*, March 13, 2008.

54. Marlow Hood, "Wall Street Underwater," *Australian Age,* March 16, 2009.

55. William Yardley, "Study Sees Climate Change Impact on Alaska," *New York Times,* June 28, 2007.

56. Alex DeMarban, "Military to Help Eroding Village Move," *Fort Mill Times*, November 1, 2008.

57. Evan Lehmann, "Washington Is Stimulating 'Underwater' Projects," *ClimateWire*, March 27, 2009.

58. Ben Cubby, "Coal Group Coy about Port Exposure to Rising Seas," *Sydney Morning Herald*, June 15, 2009.

59. Anne Moore Odell, "Risking the Weather," http://www.socialfunds.com/news/article.cgi/2398.html, October 24, 2007.

60. "Bush Tells Gulf Coast Residents to Flee 'Dangerous Storm,'" CNN.com, August 31, 2008.

61. Rie Jerichow, "Insurance Premiums Will Rise Due to Climate Change," *Times*, March 23, 2009.

62. http://thinkprogress.org/2007/04/04/insurance-losses-to-sky rocket-with-global-warming/.

63. Paul R. Epstein and Evan Mills, eds., *Climate Change Futures* (Boston: Harvard Medical School, 2005), pp. 8–9.

64. Evan Lehmann, "Texas Insurers Suffer Record Losses," *New York Times*, March 25, 2009.

65. Jerichow, "Insurance Premiums Will Rise."

66. Ibid., p. 110.

67. Jim Giles, "Climate Change to Stifle Developing Nations' Growth," *New Scientist*, January 17, 2009.

68. http://www.ramsar.org/info/values_shoreline_e.htm.

69. "Extreme Events Claims Mounting," *Australian*, November 8, 2008.

70. Lehmann, "Texas Insurers Suffer Record Losses."

71. Epstein and Mills, *Climate Change Futures*, p. 8.

72. Ron Scherer, "Beyond Gasoline," *Christian Science Monitor*, June 5, 2008.

73. John Bonfatti, "Paving Work Delayed," *Buffalo News*, August 25, 2008.

74. Christian Sanchez, "Asphalt Prices Put Damper on Road Work," *Tennessean*, December 1, 2008.

75. Rob Taylor, "Analysis—Financial Crisis Takes Toll on Australia Carbon Scheme," Reuters, October 16, 2008.

76. Danny Hakim, "Paterson Draws Fire in Shift on Emissions," *New York Times*, March 5, 2009.

77. Ariane Eunjung Cha, "China's Environmental Retreat," *Washington Post*, November 19, 2008.

78. Ralph Jennings, "Asia Construction Frenzy Needs Green Injection," Reuters, May 5, 2009.

79. Robert Collier, "Can China Go Green?" ClimateProgress.org, December 16, 2008.

80. Joe Romm, "China Announces Plan to Single-Handedly Finish Off the Climate," ClimateProgress.org, January 9, 2009.

81. Dr. Manjur Chowdury, Aedes Larval Survey Report, November 2000, http://www.geocities.com/prevent_dengue/survey.html.

82. "Dengue Fever in Latin America," *Economist*, April 19, 2007.

83. "Brazilian Military Joins Battle against Dengue Epidemic," CNN.com, April 5, 2008.

84. John Enders, "South America Hit by Dengue Fever Epidemic," *Miami Herald*, May 17, 2009.

85. "Global Warming Would Foster Spread of Dengue Fever into Some Temperate Regions," *Science Daily*, March 10, 1998.

86. Jose A. Suoya et al., "Cost of Dengue Care in Eight Countries," *American Journal of Tropical Medicine and Hygiene*, May 2009.

87. "Cost of Dengue Challenges National Economies," *Medical News Today*, July 8, 2003.

88. "Malaysia's Dengue Deaths Mount," CNN.com, August 3, 2007.

89. Kari Lydersen, "Risk of Disease Rises with Water Temperatures," *Washington Post*, October 20, 2008.

90. Barbara Fraser, "The Andes Triple Bottom Line," dailyclimate.org, May 11, 2009.

91. "Suffering the Science," Oxfam Briefing Paper, July 6, 2009, p. 23.

92. Crystal Gammon, "Changing Climate Increases West Nile Threat in U.S.," dailyclimate.org, March 20, 2009.

93. Charles Mangwiro, "Floods Wash Landmines into Mozambique's Agricultural Fields," African Eye News Service, March 3, 1999.

94. Emily Anthes, "Climate Change Takes Mental Toll," *Boston Globe*, February 9, 2009.

95. "An Even Poorer World," *New York Times*, September 2, 2008.

96. "No End to Food Shortages," *Scotsman* (Edinburgh), September 3, 2008.

97. Vera Kwakofi, "Powering Africa's Future," BBC World Service, December 7, 2008.

98. Roberts, *The End of Oil*, p. 198.

99. "A Growing Global Power Crisis," EnergyTechStocks.com, September 2, 2008.

100. Tom Athanasiou and Paul Baer, *Dead Heat* (New York: Open Media, 2002), p. 70.

101. "Bush: Kyoto Treaty Would Have Hurt Economy," Associated Press, June 30, 2005.

102. "Climate Change Affecting China," BBC News, February 6, 2007.

103. "Temperature for Beijing Hits Record for February," FinFacts.com, February 6, 2007.

104. Collier, "Can China Go Green?"

105. "Climate Change Affecting China."

106. Arthur Max, "Ex–Bad Boy China Praised at Climate Talks," Associated Press, December 2, 2008.

107. Jeffrey Ball, "The War on Carbon Heats Up Globally," *Wall Street Journal*, December 2, 2008.

108. Nita Bhalla, "Financial Crisis Sparks Concern—U.N.," Reuters, February 19, 2009.

109. James Kanter, "The Cost of Adapting to Climate Change," *New York Times*, August 28, 2009.

110. F. James Sensenbrenner, "Technology Is the Answer to Climate Change," *Wall Street Journal*, April 3, 2009.

111. Brian Faier, "Obama Urges Congress to Complete Stimulus Package," Bloomberg.com, February 4, 2009.

112. Oliver Willis, "President Obama Won't Let 'The Perfect Be the Enemy of the Good,'" oliverwillis.com, July 20, 2009.

113. Leslie Kaufman, "Disillusioned Environmentalists Turn on Obama as Compromiser," *New York Times*, July 10, 2009.

114. Josh Braun, "A Hostile Climate," *Seed*, August 2, 2006.

115. Alex Perry, "How to Prevent the Next Darfur," *Time*, April 26, 2007.

116. "A Region in Crisis," *World Ark*, September–October 2008.

117. Talal El-Atrache, "160 Syrian Villages Deserted 'Due to Climate Change,'" Agence France-Presse, June 1, 2009.

118. A. Kitoh et al., "First super-high-resolution model projection . . . ," *Hydrological Research Letters* 2 (2008): 1–4.

119. Jack Shenker, "Nile Delta: 'We Are Going Underwater,'" *Guardian*, August 21, 2009.

120. Priyanka Bhardwaj, "Destroying the Glacier to Save It," *Asia Sentinel*, March 26, 2009.

121. Perry, "How to Prevent the Next Darfur."

122. Scott Canon, "Climate Change May Lead to Violence, Experts Warn," *Houston Chronicle*, November 16, 2008.

123. Charles J. Hanley, "Mass Migration and War," Associated Press, February 21, 2009.

124. Julianne Smith and Alexander T. J. Lennon, "Climate Change Increases the Risk of Terrorism," *International Herald Tribune*, December 3, 2007.

125. "A New (Under) Class of Travelers," *Economist*, June 27, 2009.

126. Lisa Friedman, "Climate Migrants Flock to City in Bangladesh," *New York Times*, March 16, 2009.

127. Lisa Friedman, "Bangladesh: Migrant or Refugee?" *New York Times*, March 23, 2009.

128. Dean Nelson, "India Fences Off Bangladesh to Keep Out Muslim Terror," *Sunday Times* (London), November 13, 2005.

129. Canon, "Climate Change May Lead to Violence."

130. Sujoy Dhar, "Rising Sea Salinates Ganges," Reuters, February 2, 2009.

131. Kelly Hearn, "United States Military Worries About Climate Change," *Washington Times*, November 13, 2008.

132. John Broder, "Climate Change Seen as Threat," *New York Times*, August 8, 2009.

133. David Stipp, "Pentagon Says Global Warming Is a Critical National Security Issue," *Fortune*, January 26, 2004.

134. Bruce Watson, "Are We Headed for Our Own Lost Decade," WalletPop.com, February 25, 2009.

135. Michelle Higgins, "Airlines Brace for More Woes," *New York Times*, July 14, 2009.

136. Bradford Plumer, "End of Aviation," *New Republic*, August 27, 2008.

137. Richard Heinberg, "Aviation," richardheinberg.com, May 14, 2008.

138. "Colombo Express," November 4, 2005, http://www.hapag-lloyd.com/en/pr/21748.html.

139. "World's Largest Container Ship Launched," gizmag.com, July 11, 2006.

140. Emma Maersk, Wikipedia, http://en.wikipedia.org/wiki/Emma_Maersk.

141. "Shipping Containers Recycled as Houses," Pristineplanet.com, November 2006.

142. Moises Naim, "The Free Trade Paradox," *Foreign Policy*, September–October 2007.

143. David J. Lynch, "Transport Costs Could Alter World Trade," *USA Today*, August 12, 2008.

144. Jeff Rubin, "The New Inflation," *Economics and Strategy*, May 27, 2008.

145. Justin Fox, "The End of the Affair," *Time*, July 6, 2009.

146. Keith Bradsher, "Trade Talks Broke Down Over Chinese Shift on Food," *New York Times*, July 31, 2008.

147. Stephen Castle, "After Seven Years, Talks on Trade Collapse," *New York Times*, July 30, 2008.

148. "The End of Free Trade," *Wall Street Journal*, August 31, 2008.

149. "Shipping: Holed Beneath the Waterline," *Independent* (London), November 6, 2008.

150. Leo Lewis, "Merchant Fleet Left Becalmed," *Times* (London), November 13, 2008.

151. "Economy Has Boat Owners Abandoning Ship," Associated Press, November 13, 2008.

152. Michael Patterson, "Biggest Bubble of Them All," Bloomberg.com, October 24, 2008.

153. "Iran to Barter Oil for Thai Rice," nakedcapitalism.com, October 27, 2008.

154. Scott Nyquist and Jaeson Rosenfeld, "Why Energy Demand Will Rebound," forbes.com, May 27, 2009.

155. Donella H. Meadows et al., *Beyond the Limits* (White River Junction, Vt.: Chelsea Green, 1992), p. xv.

156. Ibid., p. xvi.

157. Harvey Simmons, in H. D. S. Cole et al., *Models of Doom* (New York: Universe, 1973), p. 207.

158. E. F. Schumacher, *Small Is Beautiful* (New York: Harper Perennial, 1989), p. 21.

159. Robert M. Collins, *More: The Politics of Economic Growth in the Postwar World* (New York: Oxford University Press, 2000).

160. Bill McKibben, "Crashing," *Harvard Crimson*, November 13, 1980.

161. Roberts, *The End of Oil*, p. 220.

162. Richard Douthwaite, *The Growth Illusion* (Totnes, U.K.: Green Books, 1999), p. 211.

163. Donella H. Meadows et al., *The Limits to Growth: The 30-Year Update* (White River Junction, Vt.: Chelsea Green, 2004) p. 204.

164. Bill McKibben, "Hello, I Must Be Going," *Outside*, December 1997.

165. Meadows et al., *Beyond the Limits*, p. xiv.

166. Graham Turner, "A Comparison of Limits to Growth with Thirty Years of Reality," *Global Environmental Change* 18 (2008): 397–411.

167. Peter Newcomb, "Thomas Friedman's World Is Flat Broke," VanityFair.com, November 12, 2008. General Growth Partners filed for bankruptcy on April 16, 2009.

168. Rory Carroll, "Could Climate Change?" *Guardian*, October 28, 2008.

169. Jared Diamond, *Collapse* (New York: Viking, 2005), p. 504.

170. Steven Stoll, *The Great Delusion* (New York: Hill and Wang, 2008), p. 19.

171. Robert Samuelson, "A Darker Future for United States," *Newsweek*, November 10, 2008.

172. Richard Black, "Setback for Climate Technical Fix," BBC News Web site, March 23, 2009.

173. James Hansen et al., "Target CO_2," *Open Atmospheric Science Journal* 2 (2008): 217–31.

174. Kuching Unb, "30M People in Bangladesh Need Preferential Treatment," *Daily Star* (Malaysia), October 22, 2008.

175. Richard Heinberg, "Surviving a Reduction in Social Complexity," Post Carbon Institute, December 17, 2008, http://www.postcarbon.org/surviving_reduction.

3: BACKING OFF

1. Stacy Mitchell, "Sharp Rise in Shopping Center Vacancies," *Hometown Advantage*, newrules.org, July 19, 2008, p. xv.

2. Sue Kirchoff, "Heated Day of Testimony Exposes His Idea as 'Flawed,'" *USA Today*, October 23, 2008.

3. Alex Blumberg, "Giant Pool of Money," *This American Life*, May 2, 2008.

4. Katie Zezima, "Vermont Bank Thrives While Others Cut Back," *New York Times*, November 7, 2008.

5. Philip Longman and T. A. Frank, "Too Small to Fail," *Washington Monthly*, January 2009.

6. Douglas S. Robertson et al., "Survival in the First Hours of the Cenozoic," *Geological Society of America Bulletin* 116, no. 5/6 (May–June 2004): 760.

7. Rebecca Solnit, "The Icelandic Volcano Erupts," Tomdispatch.com, February 8, 2009.

8. "No Crisis in Rural Iceland," *Iceland Review*, October 2008.

9. George Bailey Loring, *An Oration, Delivered at Lexington on April 19, 1871* (Boston, 1871).

10. John K. Robertson, "A Brief Profile of the Continental Army," http://www.revwar75.com/ob/intro.htm.

11. Merrill Jensen, *The Articles of Confederation* (Madison: University of Wisconsin Press, 1940), pp. 240–41.

12. Ibid.

13. John Fiske, *The Critical Period of American History* (Boston: Houghton, Mifflin, 1888), p. 146.

14. Ibid., p. 105.

15. Peter Onuf, *The Origins of the Federal Republic* (Philadelphia: University of Pennsylvania Press, 1983), p. 182.

16. James Madison, "The Federalist No. 10."

17. James Madison, "The Federalist No. 14," "Objections to the Proposed Constitution From Extent of Territory Answered."

18. Fiske, *The Critical Period of American History*, pp. 63–64.

19. Forrest McDonald, *States' Rights and the Union* (Lawrence: University of Kansas Press, 2000), p. 69.

20. Ibid., p. 93.

21. Ibid., p. 73.

22. Richard Parker, "Government Beyond Obama," *New York Review of Books*, March 12, 2009.

23. Kenneth Stampp, *The Causes of the Civil War* (New York: Touchstone, 1992), p. 242.

24. Richard Weingroff, "The Man Who Changed America," *Public Roads*, March–April 2003.

25. "Cowlossus of Roads," RoadsideAmerica.com.

26. Paul Light, *A Government Ill-Executed* (Cambridge: Harvard University Press, 2008), p. 15.

27. Ibid., p. 115.

28. Ouida A. Girard, *Griffin, Ghost Town in the Adirondacks and Other Tales* (self-published, 1980), p. 24.

29. William Leete Stone, *Life of Joseph Brant—Thayendanegea*, vol. 2 (New York: George Dearborn and Co., 1838), appendix, p. iv.

30. Charles Walter Brown, *Ethan Allen* (Chicago: M. A. Donohue, 1902), p. 43.

31. Onuf, *The Origins of the Federal Republic*, p. 145.

32. Ibid., p. 142.

33. "Zogby Poll Finds Nationwide Support for Secession," VermontRepublic.org, July 24, 2008.

34. Eric Kleefeld, "Rick Perry: I Have Never Advocated," Talking Points Memo, May 18, 2009.

35. Jeff Vail, "A Resilient Suburbia 4," http://www.jeffvail.net/2008/12/resilient-suburbia-4-accounting-for.html.

36. Christopher Maag, "Hints of Comeback for Nation's First Superhighway," *New York Times*, November 2, 2008.

37. David Ricardo, *On the Principles of Political Economy and Taxation*, 3rd ed. (London: John Murray, 1821), p. 143.

38. Woody Tasch, interview with the author, no date.

39. Brian Halweil, *Eat Here* (New York: W. W. Norton, 2004), p. 73.

40. Glenn Rifkin, "Making a Product and a Difference," *New York Times*, October 5, 2006.

41. Ben Block, "Local Currencies Grow During Economic Recession," Worldwatch Institute, January 6, 2009.

42. Fiona Leney, "Noted for Trust," *Financial Times*, February 28, 2009.

43. Michael Shuman, *Going Local* (New York: Routledge, 2000), p. 49.

44. Kirkpatrick Sale, *Human Scale* (New York: Perigee, 1982), p. 413.

45. Tommy Linstroth and Ryan Bell, *Local Action* (Lebanon, N.H.: University of Vermont Press, 2007), p. 26.

46. The Center for Arms Control and Non-Proliferation, "FY 2009 Military Spending Request," February 22, 2008, http://www.armscontrolcenter.org/policy/securityspending/articles/fy09_dod_request_global/.

47. National Priorities Project, "The Hidden Costs of Petroleum," October 2008.

48. Miryam Ehrlich Williamson, "Your Tax Dollars at War," www.ruralvotes.com/thebackforty/?p=196.

49. Eric Louie, "Solar Panels Are Hot for the Stealing," *Valley Times*, August 30, 2008.

50. "Game Beware," *Independent*, November 17, 2007.

51. Witold Rybcyznski, *City Life* (New York: Scribner, 1996), p. 220.

52. Rebecca Solnit, *A Paradise Built in Hell* (New York: Viking, 2009), p. 372.

53. Shuman, *Going Local*, p. 21.

54. Sharon Astyk and Aaron Newton, *A Nation of Farmers* (Gabriola Island, British Columbia: New Society, 2009), p. 70.

55. "Multiscreen Mad Men," *New York Times Magazine*, November 23, 2008.

56. Adam Gopnik, "The Fifth Blade," *New Yorker*, May 11, 2009.

4: LIGHTLY, CAREFULLY, GRACEFULLY

1. Jason Jenkins, "Soybean Celebrity," *Rural Missouri*, September 2008, p. 8.

2. Bill Donahue, "King of Bionic Ag Uses Turbocharged Seeds, Precision Chemistry, and a Little TLC," *Wired*, December 2008.

3. Ibid., p. 44.

4. Steven Stoll, *The Great Delusion* (New York: Hill and Wang, 2008), p. 175.

5. Brian Halweil, "Grain Harvests Set Records, But Supplies Still Tight," Worldwatch.org, December 12, 2007.

6. James Randerson, "Food Crisis Will Take Hold Before Climate Change, Chief Scientist Warns," *Guardian*, March 7, 2008.

7. "World Failing to Reduce Hunger," BBC, October 15, 2002.

8. Sharon Begley, "Heat Your Vegetables," *Newsweek*, May 5, 2008, p. 33.

9. Bryan Walsh, "Why Global Warming Portends a Food Crisis," *Time*, January 13, 2009.

10. Nora Schultz, "Wheat Gets Worse as CO_2 Rises," *New Scientist*, August 17, 2009.

11. Ibid.

12. Jasmin Melvin, "Climate Change Threatens African Farmland—Study," Reuters, June 2, 2009.

13. "Climate-induced Food Crisis Seen by 2100," world-science.net, January 10, 2009.

14. "Suffering the Science," Oxfam Briefing Paper, July 6, 2009, p. 36.

15. Joe Romm, "Why the 'Never Seen Before' Fargo Flooding Is Just What You'd Expect," ClimateProgress.org, March 27, 2009.

16. Begley, "Heat Your Vegetables," p. 34.

17. "Suffering the Science," p. 18.

18. John Blake, "Drought Parches Much of the U.S.," CNN.com, December 15, 2008.

19. Joe Romm, "Australia Faces the Permanent Dry," Climate Progress.org, September 6, 2007.

20. Malia Wollan, "Hundreds Protest Cuts in Water in California," *New York Times,* April 17, 2009.

21. Alana Semuels, "Despair Flows as Fields Go Dry," *Los Angeles Times*, July 6, 2009.

22. "Chu: Economic Disaster from Warming," *Los Angeles Times,* February 7, 2009.

23. "Study: Global Warming Could Boost Crop Pests," *Chicago Tribune*, December 16, 2008.

24. Richard Heinberg, *Blackout* (Gabriola Island, British Columbia: New Society, 2009), p. 52.

25. Michael Pollan, "Farmer in Chief," *New York Times Magazine*, October 9, 2008.

26. "Starving and Penniless, Ethiopian Farmers Rue Biofuel Choice," Agence France-Presse, November 5, 2008.

27. Simon Cox, "U.S. Food Supply 'Vulnerable to Attack,'" BBC, August 22, 2006.

28. Michael Moss, "Peanut Case Shows Hole in Safety Net," *New York Times*, February 9, 2009.

29. Michael Moss, "Food Companies Try, But Can't Guarantee Safety," *New York Times*, May 15, 2009.

30. Tom Gilbert, interview by author, September 2008.

31. Nancy Humphrey Case, "Where Imagination Meets Farming," *Christian Science Monitor*, February 4, 2009.

32. Andrew Meyer, interview by author, no date.

33. Jeffrey Sachs, *The End of Poverty* (New York: Penguin, 2006), p. 36.

34. Ibid., p. 37.

35. A. Duncan Brown, *Feed or Feedback: Agriculture, Population*

Dynamics and the State of the Planet (Utrecht: International Books, 2003), p. 187.

36. Ibid., p. 193

37. Dennis Avery, "Keystone Alliance Gives Credit to Farmers," *Capitol Hill Coffeehouse*, February 16, 2009.

38. Rob Hopkins, *The Transition Handbook* (White River Junction, Vt.: Chelsea Green, 2008), p. 64

39. Vandana Shiva, "Poverty and Globalization," BBC Reith Lectures 2000, no. 5, April 27, 2000.

40. Peter Rosset, "Small Is Bountiful," *Ecologist* 29, no. 8 (December 1999): 63.

41. Edward Goldsmith, "How to Feed People Under a Regime of Climate Change," *World Affairs Journal* (Winter 2003): 26.

42. Brian Halweil, *Eat Here* (New York: W. W. Norton, 2004), p. 54.

43. Jules Pretty, interview by author, October 2008.

44. "Organic Practices Could Feed Africa," *Independent*, October 22, 2008.

45. Jules Pretty, *Agri-Culture* (London: Earthscan, 2002), p. 90.

46. Ibid., p. 96.

47. Klaas Martens, interview by author, October 2008.

48. "Census Shows Number of Vermont Farms Is Increasing," *Addison Independent*, February 19, 2009.

49. Rick Callahan, "More Americans Growing Food on Small Hobby Farms," Associated Press, October 5, 2009.

50. Hopkins, *The Transition Handbook*, p. 123.

51. Peter Bunyard, in Ouida A. Girard, *Griffin, Ghost Town in the Adirondacks and Other Tales* (self-published, 1980), p. 85.

52. Food First Institute, "On the Benefits of Small Farms," February 8, 1999.

53. CSA in NYC, justfood.org/csa/press.

54. Pretty, *Agri-Culture*, p. 122.

55. Jennifer Wolcott, "In Search of the Ripe Stuff," *Christian Science Monitor*, May 14, 2003.

56. Michael Shuman, *Going Local* (New York: Routledge, 2000), p. 59.

57. Pat Murphy, *Plan C: Community Survival Strategies for Peak Oil* (Gabriola Island, British Columbia: New Society, 2008), p. 195.

58. James Maroney, letter to Senate and House Agriculture Committee, Vermont, April 15, 2009.

59. Jeff Vail, "A Resilient Suburbia," theOilDrum.com, November 24, 2008.

60. Mary MacVean, "Victory Gardens Sprout Up Again," *Los Angeles Times*, January 10, 2009.

61. Sharon Astyk and Aaron Newton, *A Nation of Farmers* (Gabriola Island, British Columbia: New Society, 2009), p. 193.

62. Ibid.

63. Herbert Girard, in Girard, *Griffin*, p. 95.

64. Oliver Schwaner-Albright, "Brooklyn's New Culinary Movement," *New York Times*, February 25, 2009.

65. Indrani Sen, "The Local Food Movement Reaches into the Breadbasket," *New York Times*, September 9, 2008.

66. Pretty, *Agri-Culture*, p. 106.

67. Shuman, *Going Local*, p. 54.

68. Felicity Lawrence, "Global Banquet Runs Out of Control," *New Agriculturist* online, May 1, 2005.

69. Shuman, *Going Local*, p. 63.

70. Dan Barber, "You Say Tomato, I Say Agricultural Disaster," *New York Times*, August 9, 2009.

71. Edward Goldsmith, "Feeding the People in an Age of Climate Change," in Girard, *Griffin*, p. 57.

72. Rosset, "Small Is Bountiful."

73. Wikipedia entry, "Hurricane Mitch."

74. Annie Shattuck, "Small Farmers Key to Combating Climate Change," commondreams.org, December 2, 2008.

75. Pretty, *Agri-Culture*, p. 186.

76. Sharon Astyk, *Depletion and Abundance: Life on the New Home Front* (Gabriola Island, British Columbia: New Society, 2008), p. 239.

77. Brian Feagans, "Drivers Follow Tanker Trucks Like Groupies," *Atlanta Journal Constitution*, September 29, 2008.

78. Shuman, *Going Local*, p. 53.

79. Carl Etnier, "Douglas Leading State Down Wrong Track," *Rutland Herald*, October 26, 2008.

80. Mayer Hillman et al., *The Suicidal Planet* (New York: Thomas Dunne Books, 2007), pp. 58–59.

81. Ibid., p. 225.

82. Hopkins, *The Transition Handbook*, p. 36.

83. Luis de Sousa, "What Is a Human Being Worth?" theOilDrum.com, July 20, 2008.

84. Michael Grunwald, "America's Untapped Energy Resource: Boosting Efficiency," *Time*, December 31, 2008.

85. Amory Lovins on Energy, CNN.com, October 16, 2008.

86. Alan S. Brown, "Amory Lovins Rethinks Our Cheaper Energy Future," Stevens Institute of Technology press release, March 11, 2009, http://www.stevens.edu/press/cgi-bin/wordpress/?p=346.

87. Nicholas Kristof, "A Liveable Shade of Green," *New York Times*, July 3, 2005.

88. John McPhee, "Coal Train," *New Yorker*, October 3, 2005.

89. Greg Pahl, *Citizen-Powered Energy Handbook* (White River Junction, Vt.: Chelsea Green, 2007), p. 283.

90. Rebecca Smith, "New Grid for Renewable Energy Could Be Costly," *Wall Street Journal*, February 6, 2009.

91. David Morris and John Farrell, "Rural Power: Community-Scaled Renewable Energy and Rural Economic Development," Institute for Local Self-Reliance, August 2008, newrules.org.

92. Ian Bowles, "Home Grown Power," *New York Times*, March 6, 2009.

93. Anya Kamenetz, "Why the Microgrid Could Be the Answer to Our Energy Crisis," *Fast Company*, July 1, 2009.

94. Scott Malone, "Pickens's Pullback Could Signal Shift in the Wind," Reuters, July 8, 2009.

95. Brown, "Amory Lovins Rethinks Our Cheaper Energy Future."

96. World Future Council, "Power to the People," November 2007.

97. Pretty, *Agri-Culture*, pp. 98–99.

98. Burlington Electric Commission, "Performance Measures Report," March 2008.

99. Pahl, *Citizen-Powered Energy Handbook*, p. 173.

100. Joe Romm, "Harvard Physicist Says Never Mind on Google Energy Use," ClimateProgress.org, January 13, 2009.

101. Martin Mittelstaedt, "Forget the Lights," (Toronto) *Globe and Mail*, March 28, 2009.

102. Robert W. McChesney and John Nichols, *Our Media, Not Theirs* (New York: Open Media, 2003), p. 76.

103. Michael Wood-Lewis, interview by author, August 2008.

104. Annie Gowen, "In Recession, Some See Burst of Neighboring," *Washington Post*, May 4, 2009.

105. Kirkpatrick Sale, *Human Scale* (New York: Perigee, 1982), p. 509.

106. Marlowe Hood, "Top UN Climate Scientist Backs Ambitious CO_2 Cuts," Agence France-Presse, August 25, 2009.

• ACKNOWLEDGMENTS •

This book required help of many kinds, not least from my friends and colleagues at Middlebury College, beginning with Ron Liebowitz and Nan Jenks Jay and including an enormous number of others, including Jon Isham, Kathy Morse, Becky Gould, Janet Wiseman, Chris Klyza, John Elder, Steve Trombulak, Pete Ryan, Mike McKenna, Jeff and Diane Munroe, and of course Chris Shaw and Sue Kavanagh. I'm also grateful to everyone who helped liaise between the college and the wonderfully supportive folks at the Hunt Alternatives Fund, including Swannee Hunt, Adria Goodson, and Caitlin Wagner. I could not have managed the writing without the mental health breaks provided by the good folks at Rikert and Blueberry Hill.

At Times Books, many thanks to my careful and engaged editor, Paul Golob, and his assistant, Kira Peikoff; to old friends like Maggie Richards and new ones like Maggie Sivon and Nicole Dewey, and to all their many gifted colleagues, including Denise Cronin, Chris O'Connell, Vicki Haire, Donna Holstein, Ashley Pattison, Jason Liebman, Lisa Fyfe, and of course their spirited leader, Dan Farley.

I've worked with one agent, the wonderful Gloria Loomis, my entire career; as usual she was smart about this project from start to finish. Thanks also to her assistant, Julia Masnik. And many thanks to people who read this manuscript and helped with powerful suggestions, especially Rebecca Solnit, Alan Weisman, Tim Flannery, and Barbara Kingsolver, who really went above and beyond the call of friendship.

In the twenty years since I wrote *The End of Nature*, reporting on global warming has gone from a small trickle to a great torrent. I am grateful to many whose labors I've drawn on here. You'll find most of them in the footnotes, but a special salute to Joe Romm and his ClimateProgress.org, to Grist.org and David Roberts, and to the participants in the Hothouse listserv. Of course, Jim Hansen is one of the heroes of this volume and of this planet, whatever you call it.

The writing of this book overlapped with more than full-time organizing for 350.org. Many foundations made that work possible, and also the incredible enthusiasm of people around the world who care about the future. Their unsolicited and unrecompensed work made October 24, 2009, a very special day and helped change the conventional wisdom, which is a hard thing to do. My seven core colleagues to whom this book is dedicated have, I hope, some small sense of just how much they mean to me.

My usual network of extended family and friends were as always invaluable—Peggy McKibben, Tom and Kristy and Ellie, the Verhoveks, the Considines, the Wilsons, the Avignons, Alden and Missy Smith, and our ever-helpful and encouraging neighbors Warren and Barry King. Andrew Gardner and Caroline Damon make life better in an endless number of ways. And so many of our Ripton friends make this work easier, especially those who have helped support the North Branch School, beginning with Tal Birdsey, Eric Warren, and Rose Messner. Thanks also to Middlebury Union High School, the Mountain School, and the SEGL school in Washington, D.C.

And of course, most of all, to Pransky, to Sophie Crane, and to Sue. I love you guys and the life we've built together.

• INDEX •

JAPAN
AIRLINES

oneworld

ABOUT THE AUTHOR

BILL MCKIBBEN is the author of more than a dozen books, including *The End of Nature*, *Enough: Staying Human in an Engineered Age*, and *Deep Economy*. A former staff writer for the *New Yorker*, he writes often for *Harper's*, *National Geographic*, and the *New York Review of Books*, among other publications. He is the founder of the environmental organizations Step It Up and 350.org, a global warming awareness campaign that in October 2009 coordinated what CNN called "the most widespread day of political action in the planet's history." He is a scholar in residence at Middlebury College and lives in Vermont with his wife, the writer Sue Halpern, and their daughter.

Books by Bill McKibben

Deep Economy—In this powerful and provocative manifesto, Bill McKibben offers the biggest challenge in a generation to the prevailing view of our economy. For the first time in human history, he observes, "more" is no longer synonymous with "better"—indeed, for many of us, they have become almost opposites. McKibben puts forward a new way to think about the things we buy, the food we eat, the energy we use, and the money that pays for it all.

Enough—McKibben turns his eye to an array of technologies that could change our relationship not with the rest of nature but with ourselves. He explores the frontiers of genetic engineering, robotics, and nanotechnology—all of which we are approaching with astonishing speed—and shows that each threatens to take us past a point of no return. This wise and eloquent book argues that we cannot forever grow in reach and power—that we must at last learn how to say, "Enough."

Fight Global Warming Now—McKibben and the Step It Up team of organizers, who coordinated a national day of rallies on April 14, 2007, provide the facts of what must change to save the climate and show how to build the fight in your community, church, or college. They describe how to launch online grassroots campaigns, generate persuasive political pressure, plan high-profile events that will draw media attention, and other effective actions. This essential book offers the blueprint for a mighty new movement against the most urgent challenge facing us today.

The Bill McKibben Reader—For a generation, Bill McKibben has been among America's most impassioned and beloved writers on our relationship to our world and our environment. Now, for the first time, the best of McKibben's essays are collected in a single volume. Whether meditating on today's golden age in radio or the natural place of biting blackflies in our lives, McKibben inspires us to become better caretakers of the Earth—and of one another.